WHAT GOD KNOWS

WHAT GOD KNOWS

TIME, ETERNITY, AND DIVINE KNOWLEDGE

Edited by

HARRY LEE POE

and

J. STANLEY MATTSON

Baylor University Press
Waco, Texas

Unless otherwise noted Scripture quotations are from the New Revised Standard Version Bible, © 1989, Division of Christian Education of the National Council of the Churches of Christ in the United States of America. Used by permission. All rights reserved.

Cover Design by Pamela Poll

Lines on pp. 32–33 are from "The Desert as Garden of Paradise." Copyright © 2002, 1989 by Adrienne Rich, from *The Fact of a Doorframe: Selected Poems 1950–2001* by Adrienne Rich. Used by permission of the author and W. W. Norton & Co., Inc.

Library of Congress Cataloging-in-Publication Data

What God knows : time, eternity, and divine knowledge / edited by Harry Lee Poe and J. Stanley Mattson.
    p. cm.
Includes bibliographical references and index.
ISBN 1-932792-12-0 (pbk. : alk. paper)
1. Physics--Religious aspects--Christianity--Congresses. 2. Time--Religious aspects--Christianity--Congresses. 3. God--Omniscience--Congresses. I. Poe, Harry Lee. II. Mattson, J. Stanley.

BL265.P4W43 2006
231.7--dc22
                        2005031958

Printed on acid-free paper.

Dedicated to the scores of volunteers who gave of themselves sacrificially and lovingly to restore The Kilns

Contents

Acknowledgments

In the summer of 2002 in the university towns of Oxford and Cambridge, the contributors to this volume presented plenary addresses at the fifth triennial C. S. Lewis Summer Institute. The two-week institute took for its theme "Time and Eternity: the Cosmic Odyssey." In addition to these papers that address concerns of science and religion, the institute included examinations of the theme from the perspectives of the arts, the humanities, and the professions. While plenary sessions occupied the mornings, focused seminars and workshops filled the afternoons, and performances by musicians, actors, and dancers enchanted the evenings. Conferees of the institute stayed in the colleges of Oxford and Cambridge, taking their meals together in the grand dining halls with their long tables that encourage leisurely eating and stimulating conversation. In addition to the intellectual, aesthetic, and social aspects of life, the institute encourages the spiritual dimension through opening and closing worship services each week, through meditations offered to open each morning plenary gathering, and through voluntary prayer gatherings before breakfast each morning.

The C. S. Lewis Summer Institute is an incubator for creativity in scholarship and the arts. The program committee plans the institute with all of its variety in order to stimulate whatever the conferees may do in the ensuing years. In addition to scholarly articles and monographs, we have seen poetry, novels, songs, and plays that the institute has inspired. Writer's block breaks open and dry seasons are showered with inspiration. It does not happen by accident. The planning process begins before the last institute is over. We meet with focus groups and rely upon the criticism and suggestions of all those who attend the institute. We are most grateful to the conferees who attended the C. S. Lewis Summer Institute in 1998, because their evaluations and recommendations helped to make the 2002 institute the success it was.

We deeply appreciate those who spent a day with us in focused conversation about how to develop the theme, whom to invite as plenary speakers, and whom to invite to lead seminars and workshops. These friends met with us in Redlands, California, at Gordon College in Massachusetts, and at The Kilns near Oxford: Grady Spires, Karen Mulder, Cindy Smith, Bob Lehnhart, Alex Metherall, Luci Shaw, Karen Walker, MariAnne Van Eerden, Ed and Carolyn Blight, Paul Sailhammer, Jim Buchholz, Kelly Monroe, John Skillen, Jim Zingarelli, Robert Hermann Peter Dawes, and Michael Cassidy.

The hard work of sifting all of the marvelous recommendations and crafting the 2002 Summer Institute fell to the program committee. The tireless and patient dedication of the committee members through several refinements of the program resulted in the outstanding group of speakers who make up the contributors to this book. We acknowledge our debt to Gayne Anacker, Nigel Goodwin, Todd Pickett, and Dallas Willard who served with us. During a time of transition in the office of the C. S. Lewis Foundation, the consistent and cooperative staff support offered by Jill Fort and Michelle Kang provided the needed attention to assure a successful planning process leading up to the institute. Jill Fort in particular has given of herself far beyond the limits of professionalism. The 2002 conference could never have been what it was without her energy and faithfulness. Sharon Helton joined the staff of the institute just one month out to give overall direction to conference services. Her cool, professional skill in managing the scores of volunteer and paid staff made the 2002 institute virtually free of problems.

The program staff who provided leadership included Ben Patterson, Robert Kaylor, Hal Bush, David Clemensen, Kate Simcoe, Dan Yanik, Kim Gilnett, Joe Dietrich, Mary Anne Poe, and Walter Mattson. Conference services staff who provided logistical support for the institute included Sharon Elkins, Kevin Sanner, Jonathan Wright, Melody Pugh, Melanie Standridge, Larry Linenschmidt, Donna McDaniel, Susan Kean, Karen Kristen, Clayton Brooks, Barabara Castillo, Marina O'Brien, Barbara Venn, Shenelle Edwards, Amber Fort Salladin, Jodi Johnson, Karen Anacker, Rachael Anacker, Melinda Douros, Lou Douros, and Aimee Beach. In addition to these, we wish to offer our thanks to the army of volunteers and work study recipients who served as drivers, office support personnel, sales assistants, journalists, faculty assistants, ushers, stage hands, medical support services personnel, head residents in colleges, Kilns tour assistants, and day trip tour guides.

In each of the great university towns, we had the cheerful and professional support of a number of companies, organizations and institutions. In Oxford, we received attentive support from the following: Robert Bristow, Burton Taylor Theatre; John Prangley, Catholic Chaplaincy; William McLean, Christ Church Cathedral School; Alexandra Webb, Christ Church; Sally Dunsmore, Conference Oxford; Becky Sigsworth, Eastgate Hotel; Louise Birt, Oxford Playhouse; Anya Herklots, St. Aldateís Church; Caroline Carpenter, St. Catherine's College; Pat Oliver, St. Columba's United Reformed Church; Jason Adams, Studley Priory, Horton-cum-Studley; Sue Waldman, The Sheldonian Theatre; The Reverend Brian Mountford, University Church of St. Mary the Virgin; Dorothy Stepney, Wesley Memorial Hall; and Helen Russell, Worcester College. In Cambridge, we relied upon the helpful support of Frankie McGhee, Cambridge Tourism; Martine Gregory-Jones, Cambridge University Botanic Gardens; Rebecca Alper-Carter, Chilford Hall, Linton; Emma Rintoul, Conference Cambridge; The Reverend Nick Woodcock and Danielle Cook, Ely Cathedral, Ely; George Brown and Elisabeth Burmeister, Faculty of Music; Caroline Choat, Fitzwilliam College; The Reverend John Binns and John Talbot, Great St. Mary's Church; Marian Martin, Lyn Saunders and Lisa Beaumont, Magdalene College; Paul Hough and Ruth Allwood, New Hall; Mary Fuller and Bob Wickett, Robinson College; and Chris Jaeger and Speaking Volume, The Mellstock Band.

Marjorie Richard in the office of College Services at Union University transcribed several tapes of plenary addresses that aided us in editing this volume. Bob Lehnhart edited several audio tapes to make transcriptions possible.

Several of our contributors had staff support that made our editing process much simpler in an age in which everyone is much busier than is good for us. We express our thanks to Alicia Sanavi and Le-Ann Little at Beeson Divinity School; Jessica Maiorana at Eastern University; and Tani Trost, Sandra Dimas, Patti Townley-Covert, and Laura Martinez at Reasons to Believe.

The editing process could have been rather anticlimactic after the experience of being together in community at the 2002 C. S. Lewis Summer Institute. Fortunately, the process of working with Carey Newman at Baylor University Press has been far from dull. Dr. Newman has the creativity and scholarly acumen that made the process of moving from the conference to the book an invigorating adventure. We are most grateful for his vision, expertise, and interest in this project. The board of trustees of the C. S. Lewis Foundation has stood behind Stan Mattson as his vision for the foundation has steadily come to pass, and we express our gratitude for this faithful support. Dr. David S. Dockery, president of Union University, has supported Hal Poe in his work with the C. S. Lewis Foundation and the foundation's C. S. Lewis Summer Institute. We gratefully acknowledge our debt to Dr. Dockery and Union University. Dr. Poe has also been enabled in his work through the generous support of Charles Colson and the board of Prison Fellowship which has funded the position Dr. Poe now holds as Charles Colson Professor of Faith and Culture. This volume represents a portion of Dr. Poe's stewardship of that trust. The science and religion portion of the C. S. Lewis Summer Institute was partially supported by a grant from the John Templeton Foundation, and we appreciate their encouragement of this exploration of the interplay of science and religion on the topic of time and eternity.

Finally, we wish to acknowledge a debt we can never repay to our wives, Jean Mattson and Mary Anne Poe, who have supported us in our work for more years than can safely be put in print. Together with our children, Todd Mattson and Elizabeth Mattson Smith, and Rebecca and Mary Ellen Poe, they encouraged us in this project, as in so many others. In the end, they are our most valued advisors and critics.

Harry Lee Poe and J. Stanley Mattson
The Kilns
Headington Quarry, Oxfordshire
August 2004

Preface

From the very outset, the mission of the C. S. Lewis Foundation has been rooted in the conviction that "something has to be done" to restore a vital, meaningful, and constructive Christian presence within the mainstream of contemporary American intellectual and cultural life. The origins of this concern were rooted in the gentle but nonetheless telling jeremiad of Harry Blamires's *The Christian Mind* (1963), in which this former student of C. S. Lewis castigated the current generation of Christian scholars for their unconscionable abandonment of any meaningful stewardship of the extraordinarily rich Christian intellectual heritage, developed over centuries of brilliant scholarship and artistry.

The pages of Christian journals and popular media of virtually all Christian traditions at the end of the last millennium were filled with expressions of dire and, one must acknowledge, legitimate concerns over any number of negative trends—social, political, economic, and spiritual. A climate of growing alienation and frustration seemed to characterize the mood of Christian laity and their leaders. With it has come a certain melancholy and palpable yearning for an idealized "Golden Age" in America's past in which the prevailing presence of prominent Christian leaders, acting upon biblically grounded guiding principles, ensured a reasonable expectation of "law and order," not to mention a public consensus on matters of "common decency."

More often than not, the enemy for many was to be found within the walls of increasingly secular classrooms. From colleges and universities to high schools and even elementary schools, a large number of faculty, waxing increasingly hostile to the faith, were openly attacking the cherished values of our society and leading our youth, and society with it, down a dark path to chaos and ultimate ruin. The fault could also be laid at the door of an increasingly emboldened and irresponsible cadre of media moguls whose primary objective seemed to be the pursuit of financial gain through the cynical exploitation of the public's fascination with the bizarre and through the disparagement of all things innocent and virtuous. To make things even worse, activist courts seemed bent upon reinforcing any and all tendencies towards enfranchising aberrant behavior as a legitimate expression of "free speech" or a manifestation of a long-vaunted American tradition of "toleration" towards all. In short, to paraphrase Bob Dylan, balladeer of the sixties, "The times, they were indeed a changin'."

As we drew ever closer to the end of the twentieth century and with it, the end of the millennium, there was a palpable sense of anxiety in the air. Y2K was upon us! Apocalyptic scenarios were emerging with ever-increasing intensity, from sources both secular and religious. Whether the source of our nightmare derived from a fear of computers, terrorists, Christ or the anti-Christ, something *big*, it seemed, was about to happen! After all, something as rare as the passing of an entire millennium simply could not be expected to occur without a suitable event to mark it. Or could it?

What was time really all about? What made this particular moment in time a matter of such import? What was the history of time? When and where and how did we begin to mark it? And what have been the consequences, both seen and unseen, of its outworking? In what sense are we creatures of time? And just how is it possible to speak meaningfully "within time" about things that are "outside of time"? These were the burning questions that engaged the attention of many as the program committee for the 2002 C. S. Lewis Summer Institute first met in 1999 to consider a theme worthy of the occasion.

We had met in similar fashion every three years since 1988, when the C. S. Lewis Foundation first launched its triennial Summer Institute in Oxford. In the years that have passed, over 2000 conferees from around the world have gathered to consider the life and legacy of C. S. Lewis and explore the larger implications of that legacy in the realms of both mind and spirit. Oxbridge 2002, the fifth such triennial conference, offered a two-week experience unlike any other. Set amid the "dreaming spires" of both Oxford and Cambridge Universities, scholars from the arts and sciences, together with leaders in ministry and from the major professions, and joined by an array of literary, visual, and performing artists, would meet together to explore the theme, "Time and Eternity: The Cosmic Odyssey."

At a time of such great social, political, and spiritual turmoil, it seemed particularly opportune to consider the broader implications of our faith in Jesus Christ, a faith that simultaneously plants us, firmly and hopefully, in both time *and* eternity. The commencement of a third Christian millennium offered a unique perspective from which to reflect upon the current situation, not only within the geopolitical realm, but also within the realm of higher education, the arts, and the culture at large. In keeping with the legacy of C. S. Lewis, we invited "mere Christians" of all traditions, and seekers of all persuasions, to join with us in pursuit of a deeper understanding of the human adventure through time, enfolded in the embrace of eternity.

A world-class faculty led our thinking and reflection in daily plenary sessions. Scientists, philosophers, and theologians explored differing conceptions of time and eternity in the light of Christian thought through the ages. Historians, artists, and cultural interpreters and commentators examined the times of our lives and their relation to eternal matters. Throughout the conference, we were also privileged to nurture the renewal of our spirits through morning meditations, daily prayer, opportunities for artistic reflection, and formal services of worship encompassing liturgy, the preached word, and communion.

The desired outcome of all this was that a baptism of spirit, mind, and imagination, would refresh and equip us to return to our respective communities and engage the culture redemptively in Christ.

The spring of 1972 marked the earliest beginning of the C. S. Lewis Foundation when a small group of Christian university and college faculty joined me for a weekend retreat at a Presbyterian Retreat Center in West Simsbury, Connecticut. As we reflected upon the current state of Christian scholarship within the mainstream of contemporary American higher education, we agreed to pray and work toward the establishment of an inter-disciplinary, faculty-governed, "mere Christian" community of higher learn-ing that would be situated on or near a major secular college or university. It was this vision that ultimately led to the founding of the C. S. Lewis College Foundation in September of 1986.

That year upon resigning my position as director of Corporate and Foundation Relations at the University of Redlands, California, I began to invest my primary energies in leading an effort to help establish a meaning-ful Christian presence at the faculty level within the mainstream of Amer-ican higher education. One year later, following extensive travel, study, institutional visits, and interviews with many educators—Christians and non-Christians alike—I was joined by a group of Christian scholars and interested supporters at St. Andrew's Priory in Valyermo, California, where for five days, we deliberated over the establishment of a Christian "Great Books" college and school of visual and performing arts, to be named in honor of C. S. Lewis. Drawing upon the inspiration of Lewis's own academic settings in Oxford and Cambridge, and trusting in God's ultimate direction and provision, we embraced a vision for the founding of a "mere Christian" college to be situated on or near a major secular univer-sity setting within the United States.

In 1984, wholly independent of this initiative, another small group of Christians under the leadership of Robert Cording of San Diego had formed a limited partnership, The Kilns Association, for the purpose of purchasing and restoring C. S. Lewis's beloved home in Oxford, England. In 1986 the partnership was dissolved and a charitable company was formed for the pur-pose of carrying out the original plans.

In March of 1988, The Kilns Association, lacking the resources needed to achieve its worthy goal, proposed that the C. S. Lewis College Foundation assume responsibility for The Kilns. After much thought and prayer, the foundation agreed to assume full ownership and financial responsibility for The Kilns with the condition that the Boards of both organizations combine their efforts in pursuit of their essentially common goals.

During the following summer, the newly combined force of the C. S. Lewis Foundation launched the first of its triennial summer institutes in Eng-land. Oxford '88, a symposium on "The Christian and the Contemporary University," was conducted at St. Hilda's College, Oxford, with over 200 in attendance. More than 300 conferees attended Oxford '91, the second triennial conference held at Keble College, Oxford, meeting on the theme "Muses Unbound: Transfiguring the Imagination." The third triennial summer institute, Cambridge '94, broke new ground on the subject of science and religion under the theme "Cosmos and Creation: Chance or Dance?" with over 400 participants returning home from Queen's College, Cambridge, with a fresh vision for the Christian's role in both the arts and sciences.

The foundation hosted over 800 conferees at the international centenary of C. S. Lewis's birth, held at both Oxford and Cambridge universities in the summer of 1998. Celebrating the life and legacy of Lewis with the theme "Loose in the Fire," based upon the biblical text of Daniel 3:25, Oxbridge '98 paid tribute to Lewis's scholarship, creativity, and courage in both living and defending the Christian faith within a university environment that was increasingly hostile toward all things Christian. The fifth triennial C. S. Lewis Summer Institute, Oxbridge 2002, explored the theme "Time and Eternity: the Cosmic Odyssey," the fruits of which are the subject of this book. At the time of this writing, we eagerly anticipate the sixth such triennial summer conference, Oxbridge 2005, centered upon the theme "Making All Things New: The Good, the True, and the Beautiful in the 21st Century," after which the Institute will convene annually, moving to Williams College in Massachusetts in 2006 and the University of San Diego in 2007 before returning to Oxford and Cambridge in 2008.

As the summer institutes took on greater substance, the foundation's faculty-led Board of Trustees resolved at its strategic planning retreat in June of 1993 that plans for the establishment of C. S. Lewis College should temporarily be put on hold in order to allocate the foundation's limited resources to the important work of leading an all-volunteer effort to restore The Kilns. To everyone's joy, the restoration was completed in time for the Oxbridge 2002 dedication of The Kilns as home to the C. S. Lewis Study Centre. The Centre now serves the needs of visiting Christian scholars and artists-in-residence throughout the academic year and provides Seminars-in-Residence for students of Lewis during the summer months.

July of 2001 marked the first offering of the now flourishing series of Summer Seminars-in-Residence at The Kilns. These week-long seminars provide an intimate group of eight readers the rare opportunity to engage Lewis's works on location in the good company of inspired seminar leaders of the stature of professors James Como, Jerry Root, Don King, and Bruce Edwards, to name but a few.

The summer institutes have also, as one might expect, contributed greatly towards the development of a lively network of Christian scholars, cutting across a broad swath of disciplinary and confessional boundaries. Responding to increasing appeals for the formation of a broadly based international network of "mere Christian" faculty and administrators serving within the world of higher education, the foundation began prayerfully to consider how it might serve such a growing and compelling need.

With the encouragement of start-up grants received from the Fieldstone Foundation, the Huston Foundation, and a third anonymous foundation source, the C. S. Lewis Foundation committed itself in 1999 to launching an exploratory faculty-forum initiative to determine whether, indeed, there was a need for such an international network of Christian faculty. The forum's purpose: to encourage greater communication, fellowship, joint research, graduate student development, and other scholarly activities among "mere Christians" and related campus ministries serving within colleges and universities the world over.

In April of 2000, over 160 faculty members gathered at the Kellogg West Conference Center on the campus of California State Polytechnic University at Pomona for an all-day faculty-forum conference for Christian faculty serving within the Southern California region. Encouraged by the overwhelmingly positive response from all in attendance, a mid-year Faculty Forum dinner was held the following November at the Hilton Hotel in Costa Mesa. Over 200 attended to hear guest speaker, James Hunter, sociologist of the University of Virginia, deliver a keynote address on the subject "Inside the Post-Modern University: Rethinking the Challenges Before Us."

The year 2001 would find us on the campus of UCLA addressing the theme, "The Christian Scholar in the Contemporary University: A Place to Stand." The occasion afforded a fitting opportunity to present John Alexander, former professor of geography at the University of Wisconsin, Madison, and past president of InterVarsity Christian Fellowship, with the very first faculty forum award for Distinguished Lifetime Contribution to Faith and Scholarship. Most recently, the forum honored the distinguished sociologist, Robert Bellah, author of *Habits of the Heart*, at its Fall 2004 conference on "Free Speech: Academic Freedom and Religious Expression," fittingly held on the campus of the University of California, Berkeley.

Responding to mounting stateside requests for all-day and weekend programming for readers of Lewis, the C. S. Lewis Foundation conducted its first city-wide, all day C. S. Lewis conference at TMI in San Antonio, Texas in 2000. This was followed by two major day-long conferences in Austin, Texas, the first at St. Edwards University in 2002, the second, addressing the subject "C. S. Lewis on Living Ethically in the Post-Enron Age," convened on the campus of the University of Texas, Austin, in April of 2004.

January of 2001 marked the first ever "Inklings Weekend Retreat" at "The Cove," the beautiful conference facilities of the Billy Graham Training Center in the Smokey Mountains near Asheville, North Carolina. Encouraged by the enthusiastic response, this weekend conference not only continues to meet annually in the Southeast but has also inspired a similar weekend conference in the Southwest. The foundation launched its first annual Southwest Regional Retreat at the Episcopal Retreat Center at Camp Allen in Navasota, Texas, in November 2003, where it has since settled, returning in 2004 and preparing for November 2005, when it will consider *The Lion, the Witch and the Wardrobe*, in anticipation of the December release of the feature film.

June 2004 witnessed the second Summer Weekend Conference in the U.S. at the University of San Diego with the theme: "Wrestling with Culture: C. S. Lewis and J. R. R. Tolkien as Spiritual Mentors." Paul Ford, Peter Kreeft, and Joseph Pearce spoke at the gathering, joined by performers Tom Key, Fernando Ortega, and Annie Herring.

As one recalls the words of Bob Dylan's prophetic lyrics of the sixties in "The Times, They Are A-Changing", one cannot but acknowledge that he was, indeed, speaking the truth. It was a truth that registered in very different ways depending upon the audience—to the young, rebelling against the old order of tradition, it came like a breath of fresh air and a call to arms; to those steeped in the ways of faith, it more often came like a dark and threatening foretaste of the coming apocalypse. Whatever the response, the remarkable words are still worth pondering. Two phrases in Dylan's lyric give voice to a serious disposition of the C. S. Lewis Foundation in all of its endeavors, not least in the course of putting together this rather extraordinary Oxbridge symposium on the subject of "Time and Eternity."

The first is the appeal to "start swimming" instead of simply complaining. Does the Christian community really have anything to say that is truly worth hearing? We can lament the prominence of secularists in the media, but where is our story? What are we actually doing that is worth reporting?

The second is the injunction to writers and prophets to "keep your eyes wide . . . and don't speak too soon." It is the occupational curse of virtually all scholars, myself included, to have an opinion on all matters and a ready disposition to defend it. It takes grace to be still, to reflect and attend to the opinions of others, affording the Holy Spirit opportunity to open one's mind and imagination to new vistas and new possibilities. As you read the following essays on "Time and Eternity," allow your mind and spirit to wander over terrain you may not have traversed before. Look! . . . Listen! . . . and then, by all means, do heed Blamire's admonition . . . and speak!

Yours for the journey,

Stan Mattson
President
C. S. Lewis Foundation

Introduction

The question of the meaning of time has become "timely" during the present transition between decades, centuries, and millennia. More important than the portrayal of this period as a cultural event, however, is the consideration of profound issues currently under examination by the faith and science communities and the inevitable cultural impact of these issues. This book addresses one of the most difficult concepts for both science and religion.

Within religion, the relationship of God to time raises many questions, not least of which includes what God can know. As physical beings we live in time, but as spiritual beings we are bound for eternity. For this reason, the theme of time and eternity has implications for every discipline and vocation.

Within science, the understanding since Einstein of the relative nature of time has been compounded in its complexity by the theoretical understanding of the multidimensional nature of time at the quantum level. Science and religion, each in its turn, have had a major impact upon the social sciences and

1

the arts which frame the cultural context in which scientists and theologians pursue their work.

Physicists are currently exploring the multidimensionality of time, whether time had a beginning, and even if time exists. Stephen Hawking has made the effort to show that the universe did not necessarily have a temporal beginning as Big Bang cosmogeny suggests, because of the significant religious implications of a Big Bang beginning of time. On the other end of time, Robert Russell is exploring the physics of the new creation. Theologians are questioning God's relationship to time, the meaning of eternity, whether God knows the future, and what the implications of a future new heaven and new Earth might mean for the existing physical realm. Can physics shed light on how God might have spoken through the prophets concerning future events? Does the faith concept of the eternal offer any insight to physics? What are the implications of this discussion for the classical understanding of God as omniscient, omnipresent, and omnipotent?

How society will view time in the future has tremendous implications for the culture as a whole. Different cultures have viewed time differently, as the Greek *chronos* (chronological time) and the Hebrew *kairos* (appropriate time) demonstrate. In the contemporary world, African culture, Asian culture, and Western culture have different concepts of time as it relates to life. The concept of time has implications for morality and commerce, politics and art. One's concept of time expresses itself in such matters as linear versus non-linear thinking.

For many centuries, theologians and scientists approached their work with the assumption that time is constant and absolute throughout the cosmos. The physics of Isaac Newton, upon which so many other disciplines depended, enshrined this assumption along with a static view of an eternal universe. When Einstein destroyed the old view of the universe, he destroyed the old notion of time with it. His theory explained that time is a dimension of the physical cosmos like space, and like space it is relative. After one hundred years in Einstein's universe, theology, philosophy, and science have not reached a consensus within their own disciplines about what time is.

Though these disciplines lack a consensus or even a prevailing view of the nature of time, an overwhelming consensus holds that the implications are strikingly important. One of the leading theological concerns of the twentieth century involved the knowledge of God. Mainline theologians argued the question in the first half of the century and evangelicals took up the debate in the last decade. The issue hinges on how God relates to the time-space continuum, but a surprising number of theologians seem ignorant of the implications of Einstein's theory. At the same time, many theologians still operate as though Newton's conception of time were authoritative. Physicists, on the other hand, are torn over whether time actually "exists" or is a mental construct. They further question whether it "flows" or exists as a block. The implications of a time-space continuum lead to a variety of alternative interpretations. Further theories, such as string theory, continue to appear with the disturbing poverty of a way to test the theories.

No single individual has the expertise to examine the breadth of the problem; therefore, this collection of essays involves a collaboration of theology, philosophy, physics, astronomy, sociology, biblical studies, and history to explore the theoretical aspects of the problem and their implications for faith. These contributing scholars will provide a basis for other disciplines to explore the implications of time and eternity for people of faith. How people understand time will have some impact on how they use the time they have.

These essays by no means exhaust the possible options currently under consideration. They do, however, suggest the kinds of questions which require attention. These essays also do not represent a consensus among themselves, but they illustrate the areas in which Christians may disagree, and they demonstrate the methodologies, thought processes, and values that form the ingredients of disagreement.

The chapters that make up this book were prepared for the 2002 C. S. Lewis Summer Institute in Oxford and Cambridge. The institute took as its theme "Time and Eternity: the Cosmic Odyssey" and was guided by Ecclesiastes 3:1-14 as a theme text. Though science and religion currently have no consensus about the meaning of time and eternity within their own spheres, much less in dialogue with one another, the contributors to this volume suggest several important proposals that may be of help in moving the discussion forward. In preparation for the institute, the contributors were asked to ponder the significant, powerful, dangerous, redemptive ways in which we can view time, eternity, and life as people seeking to imagine reality in its divine and redemptive fullness.

Chapter 1

The Problem of Time in Biblical Perspective

HARRY LEE POE

The Bible forms the background for any discussion of science and religion in the West. Whether one accepts its authority as revelation by God or regards it as a collection of culturally framed, disconnected beliefs collected over a period of centuries, the Bible forms the context for faith or skepticism. The dominant religious view in the West for the last fifteen hundred years has been Christianity. In this context, modern science as it is practiced all over the world developed. Each religious tradition that engages modern science has its own points of conflict with science that arise from each religion's understanding of the world as people experience it. Buddhism, Hinduism, and Islam, to name a few, have their own issues with science, and science likewise has issues of difference with each religion. In the West, however, one of the greatest areas of misunderstanding, but also of fruitful discussion, concerns *time*.

For centuries, the Bible served as the foundational source for the Western understanding of reality. People of learning saw no conflict between what the Bible could teach us as the revealed word of God and what the physical world could teach us as the handiwork of God. It seemed reasonable to suppose that God would be consistent in what he said and did. This situation continued for several centuries after the emergence of modern science from Copernicus onward. In conceiving the scientific method, Francis Bacon (d. 1625) observed that the obstacle to the discovery of new knowledge lay in the philosophical prejudices and presuppositions about the nature of reality through which people of learning viewed their data. Though Aristotle's philosophy had sparked an interest in the examination of the physical world during the high Middle Ages, it also provided a doctrinaire explanation for what one found in the physical world. Bacon developed the scientific method precisely to liberate the study of the physical world from tradition and the speculative metaphysics of philosophical systems, and of Aristotle's system in particular. The arrest of Galileo (d. 1642) illustrates Bacon's concern. The academic community could not tolerate his rejection of Aristotle's cosmology. The fact that his experimentation, by means of the telescope he had invented, demonstrated that the moon is not the perfect sphere Aristotle's philosophy had demanded held little weight against the ingrained philosophical view of the academy.

While the revolution in knowledge of the physical world received a great push by Bacon and Galileo, another of their contemporary scholars contributed to the revolution in biblical studies. James Ussher (d. 1656) made great strides in the critical study of the Hebrew texts of the Bible, as well as in the study of documents of the Patristic period in the early church. His study proceeded without reliance on tradition, but depended upon examination of the texts themselves. Most people know Ussher today, however, as the man who calculated the creation of the world at 4004 B.C. based on his study of the Hebrew texts. For 150 years, no one had any reason to question the calculations of such an eminent and "scientific" scholar, for Ussher embodied Bacon's scientific method. As the Protestant Archbishop of Armagh in the predominantly Catholic country of Ireland, Ussher had no difficulty dispensing with Rome's traditional understandings of patristic texts. The author of particularly pointed assaults on the Church of Rome and a well-known sympathizer with the Puritans, Ussher maintained an objective view when distinguishing genuine from spurious epistles by St. Ignatius of Antioch.

In the early nineteenth century, geologists and paleontologists made discoveries that challenged the long-established dating of the age of the world.

Rather than a six thousand year old planet, the new sciences proposed that the earth must be millions of years old. Not only did the earth appear to be older than Ussher had said, but the forms of life preserved in the fossil record suggested that the creation itself had taken more than six days. Men of science rose to the occasion. In 1813 Robert Jameson, a Scottish geologist, proposed the "Age-Day Theory" of creation whereby the days of Genesis 1 should be understood as extended periods of time or ages.[1] In 1823 William Buckland, the great Oxford geologist, proposed his famous "Gap Theory" to explain the discrepancy between the most recent discoveries of science and Ussher dating. The Gap Theory proposed a gap of extended time between Genesis 1:1 and 1:2 to account for the missing ages.[2] Buckland argued that the geological catastrophes occurred during this gap. Though Jameson and Buckland believed that some interpretations of the Bible conflicted with the science of the day, they did not repudiate the Bible. The popular imagination, however, is not so subtle.

Once the popular imagination conceived of a discrepancy between the Bible and the new knowledge obtained through scientific investigation, Darwin's concept of natural selection as an explanation for the variety of life forms found fertile ground to grow. In the face of the nineteenth-century understanding of the certainty of scientific knowledge, the Bible began to be seen in some quarters as unreliable. In the eighteenth century, the intellectual conversation between science and religion had largely focused on making a case for the validity of general revelation (nature) in comparison with the certainty of specific revelation (the Bible). By the end of the nineteenth century the very notion of revelation from God was under serious attack.

Conflicting notions of the meaning of time played a major role in the fracture between science and religion that grew acute after Darwin published *On the Origin of the Species by Means of Natural Selection* (1859). What was not apparent in the seventeenth century, the nineteenth century, or the present day is that the discrepancy lies between the *interpretation* of the physical evidence and Ussher's *interpretation* of the biblical texts. Ussher was a modern man. He helped to create the modern mindset. He was a product of the age of moveable type and, more importantly for our purposes, the mechanical clock.

Since "the dawn of time" (notice the modern way of thinking about the world), people had marked its passage through their experience of the world (the movement of the sun, moon, and stars, the crowing of the cock, the migration of birds, the color of plants). In the world of Ussher, few individuals owned a clock, but every town of any size had its public clock on a prominent tower to regulate the affairs of people. Elegant people of the modern world dine at eight o'clock. Premodern people eat when they are hungry. Modern industry performs its functions under exacting schedules and timetables for production and delivery. Traditional agriculture, however, operates under a different understanding of time. Farmers do not harvest crops on September 1, but when the crop is ripe.

The measurement of time goes back thousands of years as many cultures tracked the motion of the sun, moon, and stars; but the consciousness of measured time belonged to the elite of a culture and to the sacred ceremonies. The sundial, the hour glass, candles, water clocks, and other elaborate schemes for measuring time were tools of the elite of classical and Chinese culture, but the measurement of time did not belong to the cultures at large in the same way until the mechanical clock appeared. The Roman army used hours to divide the day, but the ancient Hebrews knew nothing of hours and minutes. The Hebrews and people like them needed no more precision than vague terms like "morning," "evening," and "the third watch of the night." Even though the watch in the night suggests a division of time, it actually involves a human activity or quality of how the time was spent.

The Greeks introduced the Western world to a concept of time that can be quantified and measured out like space. This concept of time is known by its Greek name, *chronos*, from which we derive the term *chronology*. *Chronos* time is the time of science, for it allows measurement. Without *chronos* time, scientific measurement would be difficult to conceive or develop. Chemistry and physics depend upon measurements that take place within prescribed periods of time. The Hebrew concept of time, on the other hand, has an entirely different orientation. Instead of quantity, it is concerned with quality. It is referred to as *kairos* time. *Kairos* does not concern itself so much with when or for how long something happened. The modern world has little use for this concept, though it is preserved in expressions like "A good time was had by all."

The distinction between *chronos* and *kairos* can be seen in the Bible in descriptions of the birth of Jesus. From the Jewish perspective, Jesus was born when "the days were accomplished" (Luke 2:6 AV), but from the Greek perspective, he was born during the reign of Augustus Caesar when Quirinius was first governor of Syria (Luke 2:1-2). From the Jewish perspective, the incarnation and atonement of Christ came in "the fullness of time" (Gal 4:4), but from the Greek perspective it began with his birth during the reign of Herod the Great and finished with his crucifixion under Roman governor Pontius Pilate when Herod Antipas was tetrarch of Galilee and Caiaphas was high priest (Matt 2:1; Luke 23:1, 7; John 18:13). From the *kairos* perspective, Paul could say that "at the right time Christ died for the ungodly" (Rom 5:6), but from a *chronos* perspective, Jesus suffered under Pontius Pilate (Acts 13:28). The Hebrew concept of time allowed for the same sort of dating of events in terms of the reigns of kings, such as Isaiah's vision which took place in the year that King Uzziah died (Isa 6:1), but the orientation is to mark time by significant events in the course of life. People in the modern world will still date events in this way as do the people of Louisville, Kentucky, who remember the great Ohio River flood or the tornado of 1974. The reign of a decadent king has a general dating: "Now in the eighteenth year of King Jeroboam son of Nebat, Abijam began to reign over Judah. He reigned for three years in Jerusalem" (1 Kgs 15:1-2). Ezekiel dated his call meticulously down to the day from the catastrophe of the fall of Jerusalem: "In the thirtieth year, in the fourth month on the fifth day of the month, as I was among the exiles by the river Chebar, the heavens were opened and I saw visions of God" (Ezek 1:1).

The contrast between *chronos* and *kairos* also appears in the Passion narratives of the gospels. Both Greek and Hebrew culture met in first-century Palestine. Greek culture had conquered Rome even though the Roman army had conquered Greece. The gospels mix the Hebrew concept of time with the Greek concept of time that had come into common usage in the Roman Empire in which Jesus lived and through which the Christian faith spread. The Romans divided the day into hours; Hebrew saw each day as a whole and only spoke of the quality of different aspects of the day, such as, the cool of the day or the heat of the day. By Roman reckoning, the day began at dawn, thus, Jesus was crucified at the sixth hour (noon) and remained on the cross until the ninth hour (Luke 23:44). By Jewish reckoning, however, the day began at sunset, and the day after the crucifixion was the Sabbath. So as not to defile the Sabbath with crucified bodies, those being crucified had to be killed rather than drawing out the slow death of crucifixion (John 19:31-42). It was a simple difference in understanding of the nature of a day. The Hebrews understood that the day began in darkness but was overpowered by the light. The Greeks, on the other hand, believed that the day began with light and slowly died. For the Hebrews, each day was new, but for the Greeks there was an endless cycle of death and rebirth.

Different understandings of the meaning of time also appear imbedded in the languages people use. Language reflects the thought processes and understandings of people. The Hebrew language has no tenses to its verbs. Hebrew verbs do not express past, present, or future tense. Instead, Hebrew verbs describe the quality of the action. The verbs describe completed action or incomplete action. The Hebrew language expresses its strong sense of the linear movement of time through other words or phrases; such as, *the days are coming, at that time, now, then, once,* and *when.* What does a translator do when they translate a language without tenses into a language that has tenses? Translation involves more than merely finding the word in another language that has the equivalent meaning. Language involves more than naming objects. Some two hundred or so years before the birth of Christ, Jewish scholars in Alexandria translated the Hebrew Scriptures into Greek for the Hellenistic Jewish community that had grown in Egypt after the fall of Jerusalem. This translation, known as the Septuagint, began a tradition of rendering the language of the Bible into a language that had a different concept of time. Modern translations of the Hebrew Scriptures into English continue this tradition.

THE BEGINNING

The first example of the clash between the Hebrew understanding of time and the Greek understanding appears with the first word in the Bible. English translations render it "in the beginning." A casual survey of the seemingly endless stream of new translations of the Bible illustrates how deeply rooted the Greek understanding of time affects our reading of the Scriptures. This traditional way of translating the word is followed by the Authorized Version of 1611 (King James), the New American Standard Version, the New International Version, the English Standard Version, the Jerusalem Bible, and the New Living Translation, to name but a few. Several translations add a footnote and give an alternative translation there: "When God began to create. . . ." These translations include the New English Bible, the New American Bible, the New Jewish Publication Society Version, the Revised Standard Version, and the Anchor Bible. While these translations vary greatly on how they treat texts throughout the Bible, they march together on this word. The word *bereshith* includes the prefix preposition (*be*) that is rendered in English as "in." What is missing from the word is the Hebrew definite article (*ha*).

What is the difference between "in *the* beginning" and "in beginning"? When the Hebrew scholars in Alexandria translated the Scriptures into Greek, they kept the Hebrew emphasis. Though Greek has a definite article just as Hebrew and English do, the Septuagint retained the Hebrew form with *en arche* (in beginning). Some three hundred years later, John used the same phrase as it is found in the Septuagint to begin his gospel. When Jews translated the Hebrew Scriptures into Greek two thousand years ago, they carried over the Hebrew concept of time, but when Gentiles under the influence of twenty-five-hundred years of Greek philosophy translate Hebrew, they tended to read the Greek understanding of time into the text.

The first translation of the Hebrew text by a Gentile took place when St. Jerome produced the Vulgate (c. 405), the Latin version of the Bible that served as the standard version in the West until the vernacular translations began to appear a thousand years later. The Vulgate begins *in principio* (in beginning), but the second lesson in standard Latin grammars explains that Latin has no word for "the." It does not express the indefinite or the definite article. In the late middle ages when vernacular translations of Latin texts began to appear for popular consumption, scholars had grown used to adjusting the Latin by inserting "a" or "the" where they saw fit. By the time the Bible was being translated into European languages, the translators were adjusting the text to fit the way they saw the world.

While the Greek concept of time focuses on when something happened and the measurement of the duration of action, the Hebrew concept of time focuses on the quality of the time, and therefore on the verbs. It is not incidental that all words in the Hebrew language are derived from verbs. The focal point of Genesis 1:1 is the word *bara* (create). It is a "perfect" verb, which means that it denotes completed action. The action of a perfect verb may be completed in the past, present or future. It is not unusual for the prophets of Israel to describe a future event as a past event or a completed action because it is something that God will do; therefore, it is as good as completed. The first verse of Genesis describes creation as a completed action *before* it has even taken place! The fact that God began creation means that God completes creation. It is an accomplished fact, regardless of the quantity of time involved. In the face of alternative explanations for the physical world, ranging from the nature religions of the ancient world to the philosophical explanations of the Greeks, the Hebrews affirmed that everything came from God.

Some modern translations offer the alternative translation "when God began to create" in a footnote. The problem with this translation is that it has changed the quality of action of the verb from completed action to incomplete action. The difference has a great deal to do with what kind of universe exists. The Babylonians believed that a watery chaos preceded the creation of the world we know. By changing the verb to an imperfect, we are left with a dualistic universe in which the divine exists alongside chaotic matter. This view would fit well with Aristotle, but it is precisely this view that the Genesis account of creation rejects. The proper way to translate the first word of Genesis has perplexed scholars since the Middle Ages when Aristotle first began to influence Western thought again.[3] While scholars argue over whether the absence of the definite article precludes a reference to time, we should also note that this one occurrence stands in a unique relationship to the rest of the usages of the word in Hebrew. Whatever we make of it, it is not only the first word in the text or the first word in the creation account, but the first reference to temporality in the Bible. It does not yet have the two directions of temporality (past and future) from which to identify itself. Within the Hebrew concept of time with its linear direction, people are identified by their forebears and their offspring. This is a unique moment, for nothing has gone before and nothing has yet come after.

Having affirmed the completion of creation in the first verse, the rest of the first chapter of Genesis describes the variety in the creative activity of God. Over a cycle of seven days, God creates in different ways. God calls forth the light directly, but note the quality of the verb. It is not in the imperative mood of a command: "Be light!" Instead, it is in the voluntative form: "Let there be light." God expresses his will and light results, but not through coercion or even exertion. The Genesis account lacks the exotic character of the origins mythologies of the nature religions. It involves no struggle, no combat to the death, no uncertainty about the outcome, no rivalry, no suspense. In simplistic terms, the voluntative mood has a voluntary quality to it. In addition to being in the voluntative mood, the verbs involved in the creation of light are imperfect. An imperfect verb denotes incomplete or ongoing action. A literal rendering of this passage would read, "And then God *began to say*, 'Let there *begin to be* light.' "[4] In the Hebrew mindset, creation is not a static event, but a continuing activity of God, so much so that Solomon would complain about the monotony of it all (Eccl 1:5).

The dynamic character of creation as an ongoing action continues in the description of the creative activity of God in the first chapter of Genesis. The use of the imperfect verb describes the activity across the realm of creation:

light – 1:3

the firmament – 1:6

dry land and the seas – 1:9-10

vegetation – 1:11

celestial bodies – 1:14

creatures of the water and air – 1:20

land creatures – 1:24

Interestingly enough, this verb form of continuing action also appears in relation to people (1:26-27).

Of course, the biggest conflict between the text of Genesis and modern cosmology is that Genesis claims that creation was accomplished in seven days while modern cosmology describes a process over billions of years. Again, the modern translators have imposed on the Hebrew text a Greek understanding of time. Note the time sequence described in modern translations:

The first day:	light is created and separated from darkness.
The second day:	the firmament is created to separate the elements.
The third day:	dry land appears and the elements are further separated; vegetation emerges on the land.
The fourth day:	the stars, galaxies, and other celestial bodies appear, but they are the means of separating light and darkness described in the first day.
The fifth day:	living creatures appear in the water and take to the sky.
The sixth day:	living creatures appear on land; God makes humans in his image.
The seventh day:	God stops creating.

"'In the Hebrew text, the definite article is missing again. Verse five is not "the first day." The activity of verses 3-5 occurs "one day." The firmament is not made on "*the* second day" (v. 8), but on "*a* second day." This pattern continues throughout until the appearance of land animals which takes place on "*the* sixth day." Consider the difference in meaning.[5] With the definite article, the events transpire in sequence on consecutive days, as any Westerner would expect. Without the definite article, however, the relationship takes a different shape. The relationship is based on quality of action rather than on temporal sequence.

Consider the days of a month:

July 1 is the first day.

July 2 is the second day.

July 3 is the third day.

July 4 is the fourth day.

July 5 is the fifth day.

July 6 is the sixth day.

July 7 is the seventh day.

This sequence represents the Western understanding of the sequence of days in the Hebrew text with the definite article. If we change the definite article to an indefinite article, however, we see a different pattern:

July 1 is one day.

July 14 is a second day.

July 23 is a third day.

August 6 is a fourth day.

September 18 is a fifth day.

September 19 is the sixth day.

December 10 is the seventh day.

This sequence represents one day, a second day, a third day, and so forth. The combination of the imperfect verbs with their continuing action and the indefinite ordinal numerals paints a picture of an open view of time. The time frame could be vast:

December 25, 800, is one day.

June 15, 1215, is a second day.

May 29, 1453, is a third day.

October 12, 1492, is a fourth day.

October 31, 1517, is a fifth day.

July 4, 1776, is the sixth day.

June 6, 1944, is the seventh day.

Even with the vast expanse of time between days, this sequence still follows the Western pattern of thought.

The pattern of Genesis, with its concern for quality of action, follows a different sequence. The numbers certainly imply sequence, but a second day need not come chronologically before a third day. These can be days out of order since they are not defined by a definite article. Thus, the chronological sequence of days could be:

July 1 is one day.

July 4 is a second day.

July 2 is a third day.

July 5 is a fourth day.

July 3 is a fifth day.

July 6 is the sixth day.

July 7 is the seventh day.

In fact, this sequence seems to be the pattern of action in Genesis. Light is created on one day and ornamented later (a fourth day). Water is created on a second day and ornamented later (a fifth day). Dry land is created on a third day and ornamented later (the sixth day). On the seventh day God rests because creation is complete. In the prophetic sense, the seventh day could even lie in the future, such as the expected Day of the Lord, because of the sudden change from indicative to intensive mood. God does not simply begin to rest, but intensively begins to rest. The book of Hebrews suggests that entering into God's rest is an eschatological event (Heb 3:7–4:11).

The Representational Quality of Numbers

At this point, a word must be said about the significance of the number seven in Hebrew thought. Seven is a symbolic number that denotes completeness or totality. It does not mean what a mathematician means by the number seven in many cases, but it can mean the sum of three plus four. Whenever the number seven appears in a biblical text, the reader should be alert to the point being made. Note the assurance God gives Cain: "Whoever kills Cain will suffer a sevenfold vengeance" (Gen 4:15). In the seventh generation from Adam, note how Lamech has corrupted the idea of vengeance (justice):

> "Adah and Zillah, hear my voice;
>> you wives of Lamech, listen to what I say:
> I have killed a man for wounding me,
>> a young man for stiking me.
> If Cain is avenged sevenfold,
>> truly Lamech seventy-sevenfold." (Gen 4:23-24)

In the span of a few verses, we can see that full, complete, or perfect vengeance (justice) is represented by seven. The flowering of human corruption has reached fullness, completion, or perfection by the seventh generation. The full, complete, or perfect corruption of divine justice and the appropriation to humanity of the divine prerogative is represented by a seventy-seven fold exertion of violence. Compare Lamech's view of vengeance with the teachings of Jesus on forgiveness:

> Then Peter came to Jesus and said to him, "Lord, if another member of the church sins against me, how often should I forgive? As many as seven times?"
>> Jesus said to him, "Not seven times, but, I tell you, seventy-seven times."
>> (Matt 18:21-22)

In the Hebrew mind, the number seven represented a quality of perfection about whatever it modified, rather than representing merely a quantity. By the first century, this old mindset was already breaking down under the influence of the succession of Babylonian, Persian, Greek, and Roman captivities. Peter retained the old form of speech, but he seems to have given the form a quantitative meaning. If he forgave his brother seven times, by count, then he had satisfied all that could be reasonably expected of him. Jesus counters by retaining the old understanding of perfection. Seventy-seven times of forgiveness does not represent a quantity, but a quality of life. Jesus exhorts Peter to lead a life of forgiveness. It will not end at seventy-seven or seventy-seven thousand.

The use of the number seven appears throughout the Hebrew and Christian Scriptures. It moved from the symbolic to the concrete in the ritual ceremonies of the law in which activities take place over a period of seven days (Lev 8:33), or some action takes place seven times (Lev 4:6), or items are grouped by sevens (Lev 23:18). The ceremonial lampstand of the tabernacle had three branches on either side of the central lampstand (Exod 37:17-24). This seven-cupped lampstand, or *menorah*, visually represents the faith in Yahweh who created all things. Notice how the three branches on each side of the central light correspond to the days of creation and how day one on one side is balanced by day four on the other side (Figure 1.1). Day two on one side balances day five on the other side. Day three on one side balances day six on the other side. The *menorah* visually represents the parallelism of Hebrew poetic thought in sacred worship of the God of Creation.

Figure 1.1

The End

While the opening of Genesis poses one set of problems for the modern notion of time, the book of Revelation poses another set of problems related to time at the end of the Bible. Revelation presents a picture of the end of this world. It contains many references to time and periods of time. Just as Archbishop Ussher introduced a modern scientific paradigm for calculating the beginning of the world, a variety of attempts have been made to calculate the end of the world based on the references to time in Revelation. These attempts have been undergirded by the assumption that the Bible works like a mathematical formula in which it is only necessary to substitute the correct starting date in order to calculate correctly the date of the second coming of Christ.

The second coming has always been an article of faith for Christian believers, just as a form of the day of judgment is found in Judaism and Islam. In the early church, however, the Hebrew mindset in which the imagery of Revelation is framed did not survive the first century. Christianity soon became a Gentile religion with a Greek mindset prone to calculation. In the second century, a heightened expectation of the return of Christ occurred when Montanus, a former priest of the oracle of Cybele, announced that it had been revealed to him that Christ would return to Phrygia where Montanus lived (c. 170). Montanus gathered a great following that stressed asceticism, ecstatic tongues, and prophetic utterances. The movement continued for several centuries after the death of Montanus, but Christ never came to Phrygia. Montanus failed the test of a prophet. The return of Christ would reemerge as a major motif of faith during times of great persecution or calamity throughout the ancient and medieval period and into modern times.

In 1453 when the Turks captured the great city of Constantinople and brought the last of the Roman Empire to an end, the Eastern Church faced such a calamity. They had never known a time when church and empire had not existed together. The beginning of the end was signaled by the fall of the holy city of Constantinople, where God's chosen servant sat upon the imperial throne. Likewise, the Russian Church calculated that Christ would return in 1492, based on their interpretation of time in Revelation. So confident were they that they did not bother to calculate the church calendar for 1492. The failure of Christ to appear created another crisis until the Russians decided that, just as God had moved the seat of the empire from Rome to Constantinople, he had now moved the seat of the empire to Moscow. The Grand Duke assumed the ancient title of Caesar, which in Russian is Czar, and Moscow has been known ever since as the third Rome.

Savonarola (d. 1498) convinced the citizens of Florence that the return of Christ was just around the corner, and the citizens responded by driving the ruling Medici family out of town. During the Reformation, the Anabaptists of Münster took over their city in 1534 and set up an apocalyptic government in expectation of the immediate return of Christ. During the English civil war and interregnum, a millenarian movement known as the Fifth Monarchy expected to usher in the return of Christ by ending the remnants of the old fourth monarchy. Their attempted insurrection in 1662 following the restoration of Charles II only resulted in the enactment of the fierce measures of persecution against nonconformists that would continue in force until the Reform Bill of 1832.

Religious sects since the Enlightenment, like the Shakers, the Mormons, and the Jehovah's Witnesses, have all had a strong element of apocalyptic expectation. From time to time, small sects grow violent, like the Jim Jones cult. The Dispensational Movement of the late nineteenth century swept across the boundaries of the American denominations that came from England, primarily the Presbyterians, the Congregationalists, the Methodists, and the Baptists. Smaller emigrant evangelical denominations also were swept up in the movement that sought to understand God's timetable for the end of the world. The most important single instrument in the spread of Dispensationalism was the Scofield Reference Bible. In his introduction, C. I. Scofield complained that the old system of reference was "unscientific."[6]

In his widely distributed booklet, *88 Reasons Why the Rapture Will Be in 1988*, Edgar C. Whisenant went to great pains to insure that both the Jewish lunar calendar and the Gregorian solar calendar were taken into consideration and that all "mathematical calculations . . . fit perfectly together."[7] One passage from his booklet illustrates how Whisenant's modern worldview dominates his reading of Scripture:

> God used the Seven Feasts of Israel to tie together the 69th and the 70th week of Daniel. The 69th and the 70th weeks of Daniel are 1,958 years apart (covering the Church Age).
>
> The 69th week of Daniel ended 6 April 30 A.D. at the closing of Jesus' tomb. The 70th week of Daniel starts with the Day of Atonement 1988 when Antichrist signs the Seven-Year Peace Pact with Israel on 21 Sept. 1988, and the 70th week of Daniel ends seven Jewish years later on the Day of Atonement 1995, at the battle of Armageddon, 4 Oct. 1995, thus lasting seven Jewish years. (Note: The next paragraph is extremely important.)
>
> Lunar dates of the last three Feasts of Israel for the years 1988 through 1995 provide the beginning and the ending dates for the count of days given by God in Ezekiel, Daniel, and Revelation and verify beyond any reasonable doubt that from the Day of Atonement 1988 through the Day of Atonement 1995 is the 70th week of Daniel. (This single fact is the unchallengeable proof that this book is correct and true.)[8]

Whisenant and others who try to read Revelation (together with Daniel and Ezekiel) as a scientific text with his understanding of time, must stop the clock and restart it to make their calculations come out right. Whisenant stops his clock at the resurrection of Christ and restarts it with the establishment of the modern state of Israel.

For centuries, people have imposed their contemporary scientific under-standings upon the text of Revelation (and Daniel) to interpret the meaning of time. If the representation of the creation of the world in a sevenfold cycle has profound implications for the interpretation of the concept of time in Genesis, the presence of the sevenfold cycle in Revelation cannot be stressed enough either. In Genesis, the "days" are not necessarily periods of sequential time or periods of the same duration. In Revelation, the multiplication of sevenfold events, activities, and periods of time amplifies the need to understand Revela-tion from a Hebrew, rather than a Greek, perspective.

The Greek desire to measure and divide into equal intervals has led to manifold misinterpretations of Revelation over the centuries. John piles up the clues, however, that Revelation expresses a Hebrew understanding of time and numerology. Note the repetition of "seven" in the opening verses of Revelation:

seven churches – 1:4

seven spirits – 1:4

seven lampstands – 1:12

seven stars – 1:16.

We know there were more than seven churches in the province of Asia— Colosse for instance—but seven represents the complete church in Asia. Chris-tians have a difficult enough time explaining God as Father, Son, and Spirit without discovering there is not one, but seven Spirits of God (1:4). Again, the use of the number seven conveys the completeness of the Holy Spirit of God. The seven candlesticks are not lampstands, but the churches of Christ (1:20) while the seven stars are not stars, but the angels of the churches (1:20).

For those who try to calculate the timing of events in Revelation, three series of seven events take center stage. These events are the opening of seven seals (6:1-17, 8:1-5), the blowing of seven trumpets (8:6–9:21, 11:15-19), and the pouring out of seven bowls of wrath (16:1-21). Instead of three chrono-logically sequential series of events, these three series probably represent only one series that has been presented three times in the Trinitarian formula.[9] The Synoptic Gospels contain a discourse by Jesus to his disciples prompted

The Little Apocalypse	Seven Seals	Seven Trumpets	Seven Bowls
1. War and rumors of war Mt. 24:6; Luke 21:9; Mk 13:7	1. Conquest Rev. 6:1-2	1. Hail, fire, blood on earth—1/3 of earth burned Rev. 8:7	1. Bowl poured on earth—sores Rev. 16:2
2. Nation will rise against nation Mt. 24:7; Mk. 13:8; Lk 21:10	2. Peace taken away Rev. 6:3-4	2. Blazing Mountain in the seas—1/3 of seas turned to blood Rev. 8:8	2. Bowl poured on the seas—turned to blood Rev. 16:3
3. Earthquakes/Famine Mt. 24:7; Mk. 13:8; Lk. 21:11	3. Famine/Earthquakes Rev. 6:5-6	3. Blazing star on rivers and fountains—1/3 of waters polluted Rev. 8:10	3. Bowl poured in the rivers and fountains—turned to blood Rev. 16:4
4. Famine/Pestilence Lk. 21:11	4. Famine/Pestilence Rev. 6:7-8	4. Sun, moon, stars struck 1/3 reduction of light Rev. 8:12	4. Bowl poured on sun—scorches people Rev. 16:8
5. Persecution Mt. 24:9-14; Mk. 13:9-13; Lk. 21:12-19	5. Persecution Rev. 6:9-11	5. The Abyss is opened Rev. 9:1-3	5. Bowl poured on throne of the beast—his Kingdom plunged into darkness Rev. 16:10
6. Signs in the sun, moon, and stars Mt 24:26-29; Mk. 13:24; Lk. 21:25-26	6. Signs in the heavens Rev. 6:12-17	6. Four angels at the Euphrates are released for war Rev. 9:13	6. Bowl is poured on the Euphrates—way open for armies Rev. 16:12
7. Son of Man coming in a cloud with power and great glory Mt. 24:30-31; Mk. 13:26-27; Lk. 21:27-28	7. Fire from the altar thrown onto the earth—peals of thunder, rumblings, flashes of lightning, an earthquake Rev. 8:1-5	7. Loud voices declare "The Kingdom of the world has become The Kingdom of our Lord and of his Christ. And he will reign for ever and ever."—flashes of lightning, rumblings, peals of thunder, an earthquake, a hailstorm Rev. 11:15-19	7. A Loud voice from the Temple declared, "it is done."—flashes of lightning, rumblings, peals of thunder, a sever earthquake, hailstorm Rev. 16:17-21

Figure 1.2

A Comparison of the Little Apocalypse with the Seven Seals, Seven Trumpets, and Seven Bowls by their request to know the sign of "the end of the age" (Matt 24:3). This discourse is often referred to as the "little apocalypse" (Matt 24:4-31; Mk. 13:3-27; Luke 21:7-27). Jesus lays out a series of things that will characterize the present age until the end.[10] His series compares remarkably with the descriptions of the seven seals, the seven trumpets, and the seven bowls (Figure 1.2).

One may argue legitimately that the descriptions vary. They are not at all identical. This is an observation in keeping with the Greek mindset. Within the Hebrew mindset, however, a point is made by saying the same thing two or more times in different ways. Jesus explained lostness in terms of a lost sheep, a lost coin, and a lost son. The descriptions are quite different, but they make the same point. We find this method throughout the teachings of Jesus on the kingdom, and throughout the Hebrew Scriptures. Hebrew parallelism depends upon saying the same thing different ways:

> Create in me a clean heart, O God,
> And renew a right spirit within me. (Ps 51:10 AV)

In this famous psalm, *create* and *renew* stand in parallel or equivalent relationship to one another, as do *clean heart* and *right spirit*. In the same way, Revelation reiterates its message with a variety of images.

The little apocalypse is a straightforward narrative, but Revelation is full of the imagery of Daniel, Ezekiel, and the apocalyptic tradition of the intertestamental period. Despite its straight narrative form, the little apocalypse is not chronologically sequential. Just as we saw the creative activity of "the" fourth day linked to "the" first day in Genesis, Jesus does not mention everything in chronological order. Though persecution is mentioned fifth, Jesus says that the persecutions will precede everything else ("But before all this . . ." Luke 21:12). In the same way, the seals, the trumpets, and the bowls are interrupted with interludes out of sequence. John sees things, but the things that he sees do not necessarily have a sequential correlation. He sees them without temporal reference. He sees the 144,000 and the multitude clothed in white (Rev 7). He sees the two witnesses (Rev 11:1-14). He sees the woman and the dragon (Rev 12). He sees one beast come out of the sea and another come out of the earth (Rev 13). He sees the 144,000 with the Lamb and the harvest of the earth (Rev 14). He sees the whore of Babylon and her fall (Rev 17–18). He hears the great hallelujah and sees the victory of the white rider (Rev 19). He sees the thousand year reign and the last judgment (Rev 20). Finally, he sees the new heaven and new earth (Rev 21–22). The visions he sees in the interludes between seals, trumpets, and bowls, appear to amplify the significance of what happens within the cycles of seven, rather than standing as events in addition to the cycles. Time has a direction and a goal, but the meaning of time is not found in chronological sequence. Still, the incarnation, the atonement, and the resurrection of Christ all occur in the middle of time from a theological perspective, and represent the goal.

Just as Genesis speaks of creation in terms of days, Revelation sets the events leading up to the end in the context of days or periods of time. The two witnesses prophesy for 1260 days (Rev 11:3). The woman who flees the dragon is nourished in the wilderness for 1260 days (Rev 12:6). The Gentiles trample the holy city for fortyëtwo months (Rev 11:2). The Beast exercises authority for fortyëtwo months (Rev 13:5). The two dead witnesses lie in the street for three and a half days (Rev 11:9). After the dragon is cast down from heaven, the woman is nourished in the wilderness for "time, times, and half a time" (Rev 12:14).

People search in vain to establish the "correct" chronology of Revelation because it does not contain a chronology. Instead, it has a point of reference for the final goal toward which all of creation is moving: the new creation. Revelation gives a picture of the relationship between time and eternity without any particular correlation between the two. At one moment, John views the fall of Satan from the heavenly or spiritual realm into time and space. At another moment, he views the new creation. From the divine perspective that John is allowed to see, there is no particular moment in time and space that might be called contemporary or simultaneous with God. People organize their lives, their thoughts, and their experience of knowledge and reality through the means of time and the sequence of experience. To use the terms of computer technology, this temporal organization is a "relational data base." In Revelation, however, we see an entirely different experience of awareness that grasps human affairs without respect to chronology. This all-embracing grasp corresponds to a "flat file" system of technology that makes many information-technology professionals nervous. In this environment, the program grasps the data without the need for it to be in an ordered relationship. Revelation contrasts the human chronological experience of time with the divine perspective on time that is not experienced by the same rules of logic.

John receives a glimpse of human affairs from the perspective of eternity, but through the lens of fantastic imagery. John is shown a variety of tableaus cloaked in vivid imagery. All of the colors, numbers, gems, creatures, events, and objects have a symbolic, metaphorical, sometimes allegorical, and poetic meaning; and this kind of meaning applies to conventional time construc-tions like days, months, and years as well. Both Revelation and the Genesis 1 account of creation suggest the problem of confusing the measurement of time with time itself. The measurement of time (seconds, minutes, hours, days, years) is a human construction. Just as some people measure space by feet and inches and other people measure the same space by meters and centimeters, some people measure time by the moon and some by the sun. Either one of these methods is strange when you come to think of it, but no less strange than the method of measuring time by the vibration of atoms!

At both ends of the Bible (Genesis and Revelation) time meets eternity. Just as time and space collapse at the speed of light, the meaning of time blurs at the beginning and at the end. The meaning of a day when there is no sun, no morning, and no evening loses concrete reference and must give way to poetry or the language of analogy. Unfortunately, Bacon and Ussher belonged to the generation that killed allegory. What they wanted were cold hard facts—the tree of knowledge. Einstein and Bohr belonged to the generation that killed poetry altogether, for after the Great War, poetry ceased to be a popular art form in the West.

CONCLUSION

A modern person can read the Genesis 1 account of creation and laugh in self-satisfied amusement that the creation of stars and galaxies occurs after the coalescing of earth. Our modern person may just as well laugh that Hebrew words and sentences are read from right to left or that the ancient Hebrew alphabet has no vowels! The laughter we hear is of the person who travels to another country and complains that those people do not know how to do anything right. In the modern period, religious and nonreligious people tend to read Genesis from the perspective of a scientific view of life. With the pro-found success of modern science at discovering things about the world over the last three hundred years, truth came to be understood as brute scientific fact. For something to be true, it needed to be true by the standards of scientific investigation and experimentation. In the case of the Genesis account of cre-ation, the modern person reads the text with concerns about time: the age of the universe, the age of the earth, the elapsed time between the appearance of simple life forms and more complex life forms, the proper sequence of events and their duration. We tend to regard a nonscientific view of life as ignorant. As a result, we have seen in the last hundred years a strident effort by many well-meaning believers to defend the Genesis account of creation as a scientific statement, while nonbelievers tend to dismiss the Genesis account of creation (and the rest of the Bible) because its science is "wrong." The big theological question is not *when* and *for how long* creation took place or *when* Christ shall return. Rather, the big theological question concerns *whether* creation took place and *whether* Christ shall return.

Amazingly, the Genesis account of creation can tolerate a high level of "wrong" science. The modern world commits what C. S. Lewis called "chronological snobbery" in thinking that we finally have the right understanding of virtually everything.[11] As I have demonstrated in the analysis of the actual grammar of the first chapter of Genesis and the Hebrew understanding of time, the Genesis account allows a wide variety of scientific views. It tolerated Ptolemy and Newton, and it tolerates Einstein. Because of the thrust of the imperfect verbs that describe on-going action, it even suggests a continuing active involvement of God in the creation of life since life first appeared. Am I trying to make Genesis conform to modern science? Emphatically not! That was Archbishop Ussher's mistake, and it has been the mistake of many believers in the modern period. What happens when the science changes? Stephen Hawking is desperate to find an alternative explanation to the Big Bang theory because of its theological implications.

Time has been the big problem in forcing a scientific view on Genesis, but time is also the big embarrassment to science. With the development of the atomic clock, people of the modern world have grown increasingly sophisticated and accurate in how we measure time. Unfortunately, we have grown equally unsure about what it is we are measuring, if anything. No consensus exists today in the scientific world about the nature and meaning of time.

While some worldviews regard time as cyclical in nature, the Western worldview regards time as unidirectional. It is a movement toward the future whose meaning includes change, progress, and improvement. This view comes from the biblical perspective of reality, and people in the West live their lives with this understanding solidly enthroned. Western people hold to a linear view of time even if they have no biblical faith that God created the world in the past and will one day judge it and transform it. Science depends upon this view in order to operate.

Chapter 2

St. Augustine and the Mystery of Time

TIMOTHY GEORGE

I was asked by the program committee of the C. S. Lewis Summer Institute to frame my discussion in light of the historical and classical Christian understanding of how God relates to time and eternity. Actually, what I have chosen to do is to explore this theme within only one strand of the historic Christian tradition. It is, indeed, a decisively influential strand; namely, that of St. Augustine of Hippo, who lived from 354 to 430, and to whom we are all so much in debt as believing Christians across the ages, whom the Church catholic—not simply Roman Catholic but the Church universal—remembers as *doctor gratiae* (teacher of grace).

Hans von Campenhausen once said of Augustine that he is the only church father who even today remains an intellectual power. Some of us may want to argue for a few others being included on that list, perhaps Origen, Athanasius, Tertullian, Cyprian, and the Cappadocians, to go no further. But the fact remains that Augustine's discussion of time and eternity continues to resonate, not only among theologians such as Karl Barth, Jürgen Moltmann, Colin Gunton, Robert Jensen, and many others, but also among philosophers and postmodernist thinkers including Martin Heidegger, Jean François Lyotard, and Paul Ricoeur. Augustine is still a live quantity with whom we must come to terms. Why is this? Perhaps because, like we ourselves, Augustine lived in one of those fluid, ecotonic moments of history. Do you know the word *ecotone*? It comes from biology. An ecotone is a place where two distinct ecosystems come together, such as an estuary, where the stream of a river meets the pull of the ocean. An ecotone is a place that is inherently unstable, dangerous, ever fluid, changing all the time, but also generative, capacious, exploding with new possibilities. Now to extrapolate from biology to history, to talk about an ecotonic moment of history, it is clear that Augustine lived in that kind of world, at the pivotal intersection of two historic epochs that continue to shape our world today. And it is increasingly clear that, some 1500 years later, so do we.

Augustine's world witnessed the death throes of classical antiquity on the one hand and the birth pangs of the Middle Ages on the other. These are terms that historians throw around with great flourish—classical antiquity, Middle Ages—we know what we mean. But those who lived in the midst of times or eras like this saw things quite differently. It is not as though someone woke up suddenly one morning in, say, 850 or 1002, and exclaimed, "Ah! I am a medieval man!" No, this is a perspective that we place on these eras of history. Looking back, it seems to us that we can see that something enormous was happening culturally and historically. Augustine lived at the intersection of the dying of one culture and the slow and painful birthing of another.

Rome was meant to be eternal, *Roma aeterna*, but on August 24, 410, Alaric the Visigoth sacked the city and left it in rubble. His name itself connotes harshness and terror—Alaric the Visigoth—a figure no less terrifying to Augustine's world than Osama Bin Laden is to ours. This was a shock to Augustine and to his world. In faraway Bethlehem, St. Jerome, that ascetic scholar who had translated the Holy Scriptures into Latin, was, at the moment he heard about the sack of Rome, busily working on his commentary on Ezekiel. When he heard this news in Bethlehem he put away his manuscripts and sat stupefied in silence for three days.

Why did the Roman ideal, a worldview on which an entire culture had based its values and hopes, collapse with such sudden swiftness? Why had the gods abandoned Rome? How had eternity been so brutally ravaged by time? Indeed, *quid est ergo tempus*—what, therefore, is time? This was the question

Augustine pursued. He had already explored this theme so presciently in *The Confessions*, particularly in books 10 and 11. He returned to it again in *The City of God* which he began to write as a response to the criticism against Christians related to the fall of Rome. One of the arguments was that Rome had fallen because the Christians had abandoned the pagan gods. Now that the Christians were in charge of things, evidently the gods had abandoned Rome. Augustine began to respond to that argument in his massive *The City of God*, which he wrote over a period of fifteen years from 411 until 426, four years before his death, just when a new wave of barbarians, the Vandals, were beginning their assault on his native North Africa.

THE SPIRITUALITY OF TIME

I want to retrace just a little bit of Augustine's rather complex argument about the nature of time that was prompted by this historical event. But we must first look for what motivated him at a deeper level, for there was something deeper at work than just a response to current events. This is how Augustine begins book 11 of *The Confessions*:

> Eternity belongs to you, O Lord, so surely you can neither be ignorant of what I am telling you, nor view what happens in time as though you were conditioned by time yourself? Why then, am I relating all this to you at such length? Certainly not in order to inform you. I do it to arouse my own loving devotion toward you, and that of my readers, so that together we may declare *Great is the Lord and exceedingly worthy of praise.* . . . It is out of love for loving you that I do this.[1]

You feel the devotional pull of those words, don't you? Augustine's discussion of time and eternity is not an exercise in philosophical theology; it is rather an attempt at spiritual direction. Augustine does not explore the problem of time out of what the Middle Ages would all *vana curiositas* (empty speculation), simply in order to scratch the itch of his intellectual curiosity. Rather, he does this as a personal commentary as the opening lines of *The Confessions* indicate: "You have made us for yourself, O Lord, and our hearts are restless until they find their rest in you" (1.1.1).

Again and again in *The Confessions*, as indeed throughout his entire massive corpus of some 360 separate treatises, Augustine is ever-concerned with matters of the heart. About those whose affections are set on things that are passing away, he says, "Their heart flutters about between the changes of past and future found in created things, and an empty heart it remains." Or again, "Who shall take hold of the human heart, to make it stand still, and see how eternity . . . ordains future and past times?" (11.11.13) Always it is the heart that matters. Whatever St. Augustine has to teach us about the mystery of time and eternity, it is for the purpose of pilgrimage. This is one of his great words: *peregrinatio* (pilgrimage). Pilgrimage is not the restless wandering of Odysseus, but a different kind of odyssey—a journey with a *telos*—with a goal toward that City with Foundations whose builder and maker is God. As St. Anselm would put it so famously, all of our thinking about God, about time and eternity, is a form of "faith seeking understanding" (*fides quaerens intellectum*), leading towards vision, the beatific vision St. Paul described as an intimate and perpetual knowing and seeing "face to face" (1 Cor 13:12).

One other thing Augustine never lets us forget: the proper disposition for approaching the question of time and eternity is humility. Augustine is forever confessing his ignorance and lack of understanding about so deep a mystery. The appropriate mood is interrogative, or at best subjunctive; not declarative, much less imperative. "I am asking questions, Father; not making assertions," he says (11.17.22). "To whom shall I confess my stupidity with greater profit than to you?" (11.22.28) It is a good thing to tell God every now and then just how stupid you are! Augustine does this all the time. "Even if I had the skill to master and explain everything completely," he says, "I am limited by the dripping moments of time" (11.2.2). Doubtless this phrase, "the dripping moments of time," is a reference to what was called a *clepsydra*. It was a water clock which measured time by the slow dripping of a known quantity of water—very common in Augustine's world. The dripping moments of time—drip, drip, drip.

Because the God of eternity chooses to make his home with the humble-hearted, then we must pray with Augustine as we enter this exploration:

> O Lord . . . hearken to my soul. Hear me as I cry from the depths . . . Yours is the day; yours the night, a sign from you sends minutes speeding by; spare in their fleeting course a space for us to ponder the hidden wonders of your law: shut it not against us as we knock. (11.2.3)

This approach is very much in keeping with the perspective of C. S. Lewis. In the very first lines of his little essay, "Time and Beyond Time," printed as a chapter in *Mere Christianity*, Lewis says to the reader, "if you wish, you might as well skip this chapter."[2] Lewis tells his readers that some of them may be interested in exploring the problem of time, while others may find the topic confusing and unhelpful. Next, he proceeds to discuss his understanding of time which, with certain twists and contortions, is very much from an Augustinian perspective. Then at the end of that chapter, he declares:

> This idea has helped me a good deal. If it does not help you, leave it alone. It is a "Christian idea" in the sense that great and wise Christians have held it and there is nothing in it contrary to Christianity. But it is not in the Bible or any of the creeds. You can be a perfectly good Christian without accepting it, or indeed without thinking of the matter at all.[3]

That is an interesting statement from someone who thought about it a great deal! But this is very much in keeping with what Augustine was saying as well. As we come to this topic we recognize it is a mystery, and, as Jaroslav Pelikan has wisely said, to disclose a mystery is not the same thing as to dispel it.[4] The mystery remains, however deeply we probe it and try to understand it.

The spirituality of time—that is what we have been discussing—permeates everything Augustine writes. His commentary on time is a prayer, as indeed the entire *Confessions* is a prayer. This means that everything we say and think about time and eternity we say and think in the presence of God, *coram Deo*, not as a mere intellectual exercise, not in order to out-think somebody else, or try to, but rather as a way of coming before God and offering our lives to him as a form of faith seeking understanding, leading toward vision.

With this in mind, we turn now to look at three aspects of the Augustinian perspective on time. First, I will examine the evanescence of time, by which I mean the elusive quality of time as we experience it. Secondly, I will explore time understood as a creature of God. Finally, I will consider time as the arena of redemption and of hope—time itself as something in which God has acted decisively to redeem this world—and us in it—through the life, death, and resurrection of Jesus Christ.

THE EVANESCENCE OF TIME

We will end with Christ, but that is not where Augustine begins. In that sense he is a very modern person. Nowhere, in fact, does Augustine seem more our contemporary than in his analysis of time as an elusive reality—a vanishing ephemeral moment that has no permanence or stability. The poet Schiller put it like this:

> Three-fold the stride of time from first to last.
> Loitering slow, the future creepeth.
> Arrow-swift the present sweepeth.
> And motionless forever stands the past.[5]

Augustine put it this way: "So indeed we cannot truly say that time exists except insofar as it tends to non-being." Time is something elusive that slips the more swiftly through our fingers the more we try to analyze it or even to measure it.[6] The past no longer exists. It was once ours, but it is no longer. We cannot recall it, reclaim it, reshape it, however much we may resent it, decry it, or lament it. It's gone forever.

The finality and indelibility of the past is a kind of dead weight, and it is a dead weight that supports much of the pessimism and nihilism of our own times. The American feminist poet, Adrienne Rich, expresses this pessimism and nihilism in her poem "The Desert as Garden of Paradise."

> What would it mean to think
> you are a part of a generation
> that simply must pass on?
> What would it mean to live
> in the desert, try to live
> a human life, something
> to hand on to the children
> to take up to the Land?
> What would it mean to think
> you were born in chains and only time,
> nothing you can do
> could redeem the slavery
> you were born into?
> . . .
> Miriam, Aaron, Moses
> are somewhere else, marching
> You learn to live without prophets

without legends

to live just where you are

your burning bush, your seven-branched candlestick

. . .

What's sacred is singular:

out of this dry fork, this

wreck of perspective

what's sacred tries itself

one more time.[7]

Amidst the "wreck of perspective," without hope, without heart, we must learn to live "without legends" in these destitute times. This is all there is for those "born in chains and only time," and for those who cannot be redeemed from their slavery to an irredeemable past. In the last line of the poem Rich says that we need nonetheless to try one more time—try to summon all the energy and strength we can, to make a valiant Promethean attempt, one more time, amidst the wreck of perspective. This is our world. This is the world in which many of us live and move and have our academic being.

A second example of what I call the indelibility of the past, the weight of a past that cannot be recovered, is from Nietzsche. In his *Thus Spoke Zarathustra*, in a chapter called, interestingly enough, "On Redemption," he describes his famous concept of "the will to power." For Nietzsche this is the basis, the only basis, for courage and action on the part of the *Ubermensch* (the superman, the hero). Nietzsche says:

> Powerless against that which has been done, the will is an angry spectator to all that is past. The will cannot will backwards, and cannot break time and time's covetousness. That is the will's loneliest melancholy. The "that which was" is a stone he cannot move. This alone is what revenge is—the will's ill will against time and its "it was." [8]

Nietzsche is a type of that modern person who is forever pursued by the fear of a closed door, a slammed door, the "it was." The will can do many things, but the will cannot will backwards, he says. The past is forever gone. What about the future? Well, the future is not here now. Perhaps it will be, tomorrow or this afternoon, but there is no guarantee of that. Moreover, once the future does arrive, if it ever does, it will suddenly no longer be the future but the present! Then, without even stopping to have a cup of tea, it will slide imperceptibly into the past with fleeting alacrity, and so become a part of Nietzsche's "immovable stone," the "that which was," the closed door that is finally slammed on all of our hopes and dreams.

And so the future, no less than the past, is unavailable to us. We cannot bring it within our grasp—this despite the fact that in California, and perhaps other places in the world, some people are willing to pay enormous sums to have their body medically frozen at death and stored indefinitely in an underground freezer in hopes that medical science may one day be able to thaw it out and restore it to life. Cryogenics is a whole industry based on this premise. Despite this, the future still hangs like a Damoclean sword in mid-air, filled with fear and foreboding for all of us. No less than the past, then, the future is not ours to dispose of as we would.

If the past is no longer, and the future is not yet and may never be, surely we have the present, the now. This is the fundamental insight of existentialism in all of its various forms. The present is the arena of decision, of enjoyment, of self-expression. Do it now! Buy it now! Live it now! But Augustine is very shrewd at this point. He asks, "How do we quantify the present?" How do we measure the duration of "now"?

We can talk about, as we frequently do, this "present" century. Is a century not one hundred years? If we say this present century began in the year 2000, then already more than five years have passed. They are not present, and some ninety-seven or so remain to come. They are still in the future. They are not present, so what is the present? Well, perhaps not a century. A century is a long period of time. Let us bring it down to one year, Augustine says. Perhaps, we can say one current year is "present." We have this present year, but here we are already in the seventh month. Over half the year is gone, for good or ill, and the remaining months are not here yet. They may never come. Do you see where Augustine is going? Take it down to the span of a single day, or a simple hour, or just one fleeting solitary moment—a second—the smallest moment you can imagine. If it has any duration at all, it is divisible into past and future, and hence the present is reduced more and more and more to a vanishing point. Augustine says in Latin, *nullum habet spatium* (it has no space, no room).[9] The present is forever crowded out by the past and the future which push on it from either side to the point where there is no longer any room for it.

This is very different from the Aristotelian view of time. For Aristotle, the now, the present, is a point on a line, a part of a continuum. For Augustine, the present is always fluid; it is wholly insecure. It is ecotonic: midway between the vanished past and the unknown future. It is what William James called the "specious present." There is a term in German for it, even more graphic, I think—*die zerbröckelnde Zeit* (the crumbling time). Like a piece of moist cake, you pick it up with your fingers and it crumbles before you can even put it into your mouth. The present is like this. It crumbles and disintegrates as soon as you touch it.[10]

So time is a boundary between past and future, but it is an ever-shifting boundary, like a map of the Balkans over the last two decades. The future and the past crowd out any meaningful present, until we, too, have *nullum spatium*, no space, no place, no exit. One of the most gripping images of the inherent instability of the present, and the desperation it produces in human beings, comes from the poet Friedrich Hölderlin: "But to us poor souls is given no place to rest. Harried by pain, we grope and fall blindly from hour to hour, like water dashed from cliff to cliff in lifelong insecurity."[11]

Now it is at this point in *The Confessions*, where Augustine gives this analysis of the crumblingness of time, that he makes one of his most original moves.[12] Despite the fact that time seems to be so evanescent (it is always vanishing, we can never pin it down), Augustine notes, all of us are conscious of intervals of time. We regularly compare one particular time with another. We say, "Oh, that was a long time ago," referring to some event in the distant past. Or, "That will happen right away," like the coffee we are going to experience when this talk is over. We speak of some expected occurrence. It is not here yet, but it is going to happen. Or we might say about some contemporary event such as this conference, "It is going on right now." We talk this way all the time. How can we talk this way, Augustine asks, given the ephemerality, the evanescence, the evaporating quality of time? When we talk about past, present, and future events like this, Augustine says, we are really describing three realities in the mind. The past is present to us in memory. The present is present to us in attention. The future is present to us in expectation. So time itself is a kind of tension or strain within consciousness itself. It is (he invents this term, I think) *distentio animi*. Time is a distension of the mind or the soul. That is what time is.

Before we leave this point, I want to make a few comments about this last move Augustine has made, which we frequently refer to as the psychological understanding of time. First of all, Augustine is not giving back with one hand what he has just taken away with the other. By defining time as the distension of the mind, Augustine is simply recognizing that time, at least in its fallenness, as you and I experience it in this fallen world, is the arena of disorder. It is the condition in which our soul is constantly pulled and stretched hither and yon in various directions, its moral integrity fractured by sin and death, its very existence threatened and called into question. At one point in *The Confessions*, he says, *ecce distentio est vita mea* ("Look, this distension is my life!"; 11.29.39)

It is a very interesting word, *distentio*. It can be translated variously as "strain," "anxiety," "distraction"; but it can also mean "distorted", "misshaped," "deformed." In fact, in the Middle Ages it was a word used to describe the tortures of the Inquisition—the experience of being distended, stretched out on the rack. It is precisely from this kind of torturous state that human beings need deliverance and redemption.

Another point regarding the psychological understanding of time is this: because Augustine seems to place the reality of time inside the mind, it is tempting to read him, as some scholars have, as a forerunner of Immanuel Kant, who famously distinguished between the world as it is and the world as we perceive it, the *noumenal* and the *phenomenal*, to use Kantian jargon. For Kant, time is a conceptual framework that we construct and impose on reality in order to make sense of the world. Like God or the world itself, time is a mental grid, manufactured in the mind, an ordering concept that enables us to arrange, explain, and account for that which we encounter all around us. It is very tempting to read Augustine as a proto-Kantian because Augustine seems to reduce time to a mental state, to an understanding within the mind, a distension of the mind. I think this is fundamentally a mistaken reading of Augustine.

As we shall see, Augustine has a very solid sense of time as an extra-mental reality. In his commentary on Genesis, time does not begin when human beings are created to think about it! Why, then, all this emphasis on the evanescence of time? Why this talk about *distentio*, and being stretched out to the point of nothingness? Augustine's idea, I think, is a religious one. He wants to show us that time is never at our disposal. It is never ours to claim and control and command. Friedrich Schleiermacher, who in my view got so many things wrong in theology, was never more right than when he described religion as the feeling of "absolute dependence."[13] The evanescence of time reminds us that here we have no continuing city. We are ephemeral beings, radically dependent on the God who sustains us moment by moment by his sheer favor and love. Augustine wants to show us that our hearts will never find rest in the vanishing flux of time but only in the mercy and the grace and the patience of time's creator.

TIME AS THE CREATURE OF GOD

I want to turn now to the theme of time as the creature of God. You know, sometimes our children have the best insights into the deepest mysteries of the faith. I like the story of the little girl, five years old or so, sprawled out on the kitchen floor with a crayon and pencils. The mother comes in and asks, "What are you drawing, dear?" "I'm drawing God," she says. "But no one knows what God looks like." "They will when I'm through." Already a Kantian at age five!

In dealing with the mystery of time and eternity, Augustine takes up three questions which are often asked by children, but in fact they are not childish questions at all. They are some of the deepest questions we can ask. Question 1: What was God doing before he made the world? Question 2: What did God use to make the world? Question 3: Why did God make the world in the first place? Children's questions, but not childish.

So, what was God doing before he made the world? This was, in fact, a stock question in the religious debates of the ancient world. Already in the second century after Christ, Irenaeus confronted this question in his debates with the gnostics. He refused really to deal with it. He said, "I'm not going to speculate about this." Origen in his day considered the question again. It surfaced yet again among the Manicheans, from whom, doubtless, Augustine picked it up. There was a standard joke, which Augustine told and Calvin repeated. (You know a joke has to be pretty good to last a thousand years!) The joke went like this: "What was God doing before he made the world?" Answer: "He was busy creating hell for overly curious people like you!"

Now Augustine was aware of this joke, but he knew that it was not a sufficient reply to the serious intent behind the question, and so he gave a different answer. This is what he says: It makes sense to ask what God was doing before he made the world if, and only if, both God and the world are separate items within the same temporal continuum. But they are not. God's years, unlike ours, do not come and go. They are succeeded by no yesterday, and they give way to no tomorrow. "It is not *in* time that you precede all times, O Lord. You precede all past times in the sublimity of an ever-present reality. You have made all times and are before all times."[14] *Ante omnia tempora*, with the *ante* understood in an atemporal sense. It was Boethius who defined eternity as "the all at once whole and perfect realization of unending life."[15] This unending life, of course, has no beginning. Why? Because only God can match that definition. Thus, Thomas Aquinas was right to say, as he does in the *Summa Theologica*, that eternity is nothing less than God himself.[16]

What was this eternal God doing before he made the world? On Augustine's reading there was no such "before." There was no "then" then! Eternity is the dimension of God's own life. It has no beginning and no end, no parameters or margins or boundaries outside of God himself. On the other hand, time was willed and created by God as a reality distinct from himself. In his treatment of the world, Augustine again proves to be very original in his thinking. He says that time and the world were not only created by God but that they were created together. They were cocreated, for time is coextensive with the world. This is how Augustine puts it: God created the world not *in* time but *with* time.[17] What this means is that time is not some primordial container in which certain events happen. Time is not a receptacle; it is a relationship. A number of scholars have pointed out that Augustine's remarkable intuition of the coextensiveness of time and the world, of time and space, anticipated by some 1500 years the modern theory of relativity as developed by Einstein and others. I would simply add that Augustine arrived at this not by studying the world scientifically or by thinking philosophically. He arrived at this intuition by reflecting on the basic datum of the Christian faith: the doctrine of the incarnation.[18]

I think we can deal more quickly with the next two questions. Out of what did God make the world? This is a perfectly natural question, given the assumption of classical philosophy in all its various modalities, namely, that some kind of primordial matter had always existed. Plotinus called this preexistent material *hulē*, the Greek word for "wood." It is the lowest form of finite, creaturely existence. So creation, then, is the work of some divine craftsman, Plato's demiurge, who brings order out of chaos, who sculpts and shapes the universe out of this preexisting matter. The eternal stuff was always there.

If Augustine was original in positing the cocreation of time and the world, he was entirely traditional as a Christian in affirming the doctrine of creation out of nothing—*ex nihilo*—and *de nouveau*. For him this meant that neither time nor space could constitute a first principle alongside God. Neither *chronos* nor *cosmos* is *theos*. Both time and the world are products of the creative word of God—the *logos* of God—whom the Christian church confesses to be the eternal Son of God, of the same essence as the Father, united in love with the Father and the Holy Spirit, one God forever and ever.

And the final question: Why would such a God, the God Augustine described in his treatise *On the Trinity*, as "good without quality, great without quantity . . . everywhere, yet without place, eternal without time," decide to create the world and time along with it?[19] This is an age-old question, and it is common to all three of the world's great monotheistic religions—Judaism and Islam, no less than Christianity. There is a similar kind of answer found in all three of these religious traditions to the question of why God made the world. According to an ancient Hebrew midrash, which was current in Jesus' day, God made the world because he needed to have a partner on whom he could bestow love, for love by definition requires an object.[20] Love, to be love, must be directed towards something or someone else. Islam, too, has a version of this same motif. In a well-known *hadith* or tradition, God is acknowledged as saying, "I was a hidden treasure. I wanted to be known. Hence, I created the world so I would be known."[21] The same idea also appears in popular Christian piety, as in the folk sermon that depicts God as saying, back in eternity sometime, "I'm so lonely. I think I'll create the universe so there will be something for me to love."[22]

All of these sources tell the same story. God created the world in order to fill some deep deficiency within his own being, in order to actualize some latent possibility that otherwise would not have come true. It is precisely this kind of thinking that has given rise in recent times to what we call "process philosophy" and "process theology." The doctrine of a limited God who needs the cosmos or humanity in order to actualize his own reality is the underlying assumption in most forms of process theology.

Augustine says we do not need this hypothesis, for the doctrine of the Holy Trinity gives a credible response to this question. God does not need to create the world in order to have something to love. No. God *is* love. Had he never made the world at all, he would have suffered no deficiency, nor would he ever have been any less loving than he is now. God was never lonely. He was never bored.

A contemporary British writer, Anthony Towne, in his delightful little irreverent book, *Excerpts from the Diary of the Late God*, expresses this same idea. "I am bored with it all," God says. "Here I sit. I am omniscient, I am omnipotent, I'm omnipresent, I'm divine, I'm supreme, I am ineffable, I am, in short, God.

But I am condemned to look out interminably in all directions into an impenetrable void. Small wonder that that wretched, horny-tailed ingrate walked out on me. If I only had something to do. Something creative. I am omnibored!"[23]

We do not need that hypothesis, Augustine says. God was never bored. He was never lonely. There is within the being of God an eternal effulgence of reality, a mutuality of love that expresses itself in a dynamic reciprocity of giving and receiving. Karl Barth puts it this way:

> God loves, and to do so he does not need any being distinct from his own as the object of his love. If he loves the world and us, this is a free over flowing of the love in which he is and is God, and with which he is not content, although he might be, since neither the world nor ourselves are indispensable to his love and therefore to his being. Thus the love of God is free, majestic, eternal love. It is the eternal love in whose free and non-obligatory overflowing we are loved. And it is God himself, in all the depths of his deity, who summons and impels us to love.[24]

I like to think in images. I think images are usually prior to concepts. The image that prompts this quotation from Barth, drawing on Augustine, is that of an overflowing fountain, an inexhaustible reservoir that forever overflows, over-splashes and shares with infinite generosity that which it is. The fountain over-splashes, not because it is forced to do so by some inner compulsion or deficiency, but because at the heart of God there is this infinite generosity that is love.

TIME AS THE ARENA OF REDEMPTION AND HOPE

Now I have to come to my last point: time as the arena of redemption and hope. At the heart of the Christian faith is this stupendous claim, that the eternal God of creation, the God who *is* eternity (Aquinas), has so opened himself to our creaturely existence, to our history, to our time, that he has come among us as one of us. This is the truth of the incarnation, which we might also call the Intemporation, for in Jesus Christ, God in his own being—and not as a surrogate—has come into our own world and also into our own time, and in doing so he has taken unto himself our hurt, our pain, and indeed our sin. In doing this, Jesus Christ has shown himself to be the servant of time, because he, too, was crowded out. There was *nullum spatium* for him. As Dietrich Bonhoeffer put it, he was nudged out of the world right onto a cross. He was the servant of time, but he is also the Lord of time. He is the one who redeems time and who redeems us *in* time and so makes us ready for eternity.

Better than any theologian, I think, T. S. Eliot has expressed this with a clarity that is compelling in his choruses from "The Rock":

> Then came, at a predetermined moment, a moment in time and of time,
> A moment not out of time, but in time, in what we call history, transecting,
>> bisecting the
>> world of time, a moment in time but not like a moment of time,
> A moment in time but time was made through that moment: for without the
>> meaning
>> there is no time, and that moment of time gave the meaning.[25]

I think Eliot is a good guide here to what the Christian faith has meant by the redemption of time. Let us look at three things he says in this quotation about the incarnation: 1) It was a moment in time; 2) but it was not like a moment of time; 3) and in that moment of time, Jesus Christ was bisecting the world of time.

First, let us consider that it was a moment in time. The Christian faith is clear about this. In a wonderful passage from Galatians, the Apostle Paul says, "But when the fullness of time had come, God sent his Son, born of a woman, born under the law, in order to redeem those who were under the law" (Gal 4:4-5). The Greek word for time is in the plural here: *chronoi*. Tick, tick, tick time. Drip, drip, drip time. Time, as you and I know it, is measured in seconds and hours and days and years, by calendars, clocks and sundials. Into this kind of time God sent forth his son. In the days of Caesar Augustus, when Quirinius was the governor of Syria, Jesus was born. He suffered and died under Pontius Pilate. The name Pontius Pilatus was recently discovered in Caesarea on a stone slab dating from the first century. Pilate is a very datable, if somewhat despicable, bureaucrat of the Roman Empire. It happened ". . . under Pontius Pilate." As John puts it in that marvelous prologue, "The Word became flesh" (John 1:14). The New Living Translation renders this passage, "So the Word became human being." It is not good enough. That is too weak a christology. No, the Word became *flesh*. What Word? The Word who was in the beginning with God and was God, the one who forever shared the eternal fellowship of Father and Holy Spirit, the one who was "in the bosom of the Father," as the Authorized Version puts it (John 1:18). This Word became flesh (*ho logos egeneto sarx*).

What is flesh? Flesh is that part of our human reality which in fact is most susceptible to the ravaging of time. It is flesh that suffers pain. It is flesh that contracts cancer. It is flesh that we bury in the ground. This is what the Son of God became for us, says the Christian faith. This idea was, and still is, a remarkably shocking thought. It was put forth in the early Church over against the docetists, people who said that in Jesus Christ the Word of God touched the earth much as a tangent touches a circle, but that was it. In appearance Jesus seemed to be a real human being, but in fact this was not the case. He was a phantom-like apparition, a ghost.

Against this idea, Ignatius of Antioch and the early Christian fathers say again and again, he was "truly" born of the Virgin Mary. They use the Greek adverb *alēthōs* which means "really," "truly," "surely." That is why it got into the Apostles' Creed. He was *really* crucified under Pontius Pilate. He was *really* buried. He *really*, truly, rose again.

Then along came the neoplatonists, with whom Augustine kept good company. If I were to give a critique of Augustine rather than an explanation of his thought, here is a place where Augustine might be criticized, in my opinion. He perhaps gave too much credence to the neoplatonists who opposed the eternity of God to the reality of the world in such a way that the incarnation was an impossible thought. For the neoplatonists there is no incarnation. Even though Augustine learned much from the neoplatonists, the closer he came to Christ, the thinner his neoplatonism became. It could not be otherwise for one who took seriously the central fact of the Christian faith.

In more recent times it has been deism, in all of its various forms, that the Christian faith has had to encounter. Deism's God is the God of Thomas Hardy, who once said, "God is a dreaming, dark, dumb thing that turns the handle of this idle show."[26] Hardy's God is a thing, devoid of relationship, incapable of love. It is dark, speechless, remote, obscure. Yes, this is a hideous caricature of the real God; but it is a caricature that is widely accepted in today's world, and it is the root of much modern atheism. The Christian faith says that God does not hold *eternity* unto himself in such a way that it must forever be opposed to *time* as human beings are able to know and experience it. In a memorable metaphor, Barth declares that God takes on time for his own garment, even his own body.[27] There is a christological and a pastoral meaning to this, as expressed in Hebrews: "For we do not have a high priest who is unable to sympathize with our weakness, but we have one who in every respect was been tested as we are, yet without sin" (Heb 4:15). It was a moment in time.

Second, Eliot also says that while it was a moment in time, and of time, yet it was "not like a moment in time." You have to get this "not like-ness" in your mind. Not like. Remember what was said: He taught them as one who had authority, and not like the scribes, and the Pharisees (Matt 7:29). If Jesus were just another wise rabbi, or a cynic sage, or a charismatic healer who flourished on the edges of heretical Second Temple Judaism, then he would be of no help to anyone with AIDS, or depression, or loneliness. It was a moment in time and of time, and yet "not like." Jesus is not Socrates with a Jewish accent. He is not Plato with a beard. We will not get eternity by turning up the volume and just saying "time" in a louder and louder voice.

So the Nicene Creed confesses that Jesus Christ is of the same essence as the Father: Light of Light, very God of very God, begotten, not made. And this is the one who for us time-bound and destined-to-die creatures (who nevertheless carry eternity about in our hearts, our restless hearts), came down from heaven.

It happened in time, Eliot says, but it was not like time. And, third, it also happened in time in such a way that what we call history was torn—bisected. By that bisection, history has been given a meaning, a direction that it never had before.

In his chapter of this collection, Tony Campolo talks about the cyclical view of history, the image of history as a great cycle turning around and around. We associate this view of history with Eastern religions, and we forget that Christianity is also an Eastern religion, that it emerged, along with Judaism, in a context in which the myth of the eternal return was the dominant assumption of the age. When Paul, in Athens, preached Jesus and the resurrection, the Athenians thought he was talking about a male god, Jesus, and his female consort, Anastasis (the Greek word we translate as "resurrection," Acts 17:18). Resurrection was a part of the pantheon of dying and rising saviors celebrated in the mystery religions and, with much more sophisticated language, in the philosophy of Porphyry and Plotinus as well.

This idea of eternal recurrence also permeates the postmodern consciousness. Nietzsche said:

> How if some day or night a demon were to sneak after you into your loneliness and say, "This life as you now live it and have lived it, you will have to live

once more and innumerable times more and there will be nothing new in it, but every pain and every joy and every thought and sigh and everything immeasurably small or great in your life must return to you all in the same succession and sequence even this spider and this moonlight between the trees and even this moment and I myself. The eternal hourglass of existence is turned over and over, and you with it, a grain of dust." Would you not throw yourself down and gnash your teeth and curse the demon? Or did you once experience a tremendous moment when you would have answered him, "you are a god and never have I heard anything more godly."[28]

Nietzsche says that those who cling to the old outmoded residue of the Christian myth, who still believe in love, meaning, truth, beauty, goodness, are weaklings; they are the ones who gnash their teeth. It is the *Ubermensch* who says to the demon "Ah! You are a god." But against Nietzsche and against every form of eternal recurrence the Christian faith answers one word: *epaphax*—once and for all. There is a German word that says this much better than we do in English: *Einmaligkeit*—the once-and-for-all-ness of the Christian faith.

That is why the Christians argued so furiously about the date of Easter. What an arcane theological debate! We hardly understand what all the fuss was about anymore. Over in Ireland the monks, the Celtic monks, followed a dating of Easter very different from that promulgated from Canterbury and the Roman tradition in England, and this led to a great clash. King Oswiu of Northumbria was converted by the monks up in Lindisfarne and so he followed the Celtic dating of Easter, but his wife, Queen Eanfled, was converted by the priests from Kent, and she followed the Roman calculations. The Venerable Bede says, "Such was the confusion in those days that Easter was sometimes kept twice in one year. So that when the king had ended Lent and was keeping Easter, the queen and her attendants were still fasting and keeping Palm Sunday."[29] That will not do! This was not only a matter of domestic tranquility, it was also an issue of theological profundity: the once-and-for-all-ness of God's salvific work in Christ is at stake. We cannot have Jesus dying and rising again two or three times a year! Then we would be back into the mystery religions—the dying-and-rising-God stuff.

With this in mind, the Venerable Bede introduced our present system of accounting for time, *anno Domini* (in the year of our Lord). Now today it is not fashionable to talk about A.D. and B.C. in academic circles. We like to speak of C.E. and B.C.E., the "common era" and "before the common era." But it is still the incarnation that is the basis for this reckoning! And the year, which now begins on January 1, used to begin on March 25. Did you know that? It was not really until the late seventeenth century that the idea of starting the new year on January 1 began. That was a Roman custom; nothing Christian about it. That is why New Year's Day is not a great Christian festival. Why March 25? Because these old monks calculated that Jesus was born on December 25, and knowing a little bit about human propagation, they figured that nine months earlier the Annunciation had taken place. That was when the new era really started! Thus, the first day of the new year in Britain, March 25, was called "Lady Day" after the Blessed Virgin Mary.

A moment in time. A moment not out of time, but in time, in what we call history; tearing, bisecting the world of time. Without the meaning, there is no time, and that moment of time gave the meaning. What does this mean now for how we should live?

Conclusion

Christians are those who live in time as the ones who belong to eternity. The Christian attitude toward time is neither panic nor indifference, but hope—the hope that radiates from a messy manger, a ruddy tree, and a tomb no longer filled. The life, death, and resurrection of Jesus Christ means that God takes time, and has time, for us. Because of this, we can see time not as a threat but as a gift—a precious gift created by God, willed by God, and given by God for us. Christians are those who know that time and this world do not terminate upon themselves; they are penultimate realities that can never satisfy the deepest longing of the human heart, the restlessness. And so we live in this world ambiguously, as those who belong ultimately to another world (cf. 1 Cor 7:29-31).

My friend, Mark Noll, has said some very critical things about the old gospel song that implores,

> Turn your eyes upon Jesus,
> Look full in his wonderful face,
> And the things of the Earth will grow strangely dim
> In the light of His glory and grace.[30]

Mark Noll says, "To know God better would make our vision of the world clearer."[31] We know what he is talking about. He is protesting, and it is a valid protest, against the kind of gnostic pietism and withdrawal from the world that is so deeply rooted in evangelical culture. Yet there is a sense in which we can properly say what that song says. "Turn your eyes upon Jesus. Look full in his wonderful face, and the things of earth *will* grow strangely dim." Because there is another danger. In addition to the threat of gnostic quietism, there is the danger of idolizing this world, the danger of becoming so cozy and accommodated to the culture that we forget that this world is not our ultimate home, that we are probationers for eternity. Eternal life is not simply more and more of the same old stuff; it is God's life—God's life that comes to us as a free gift. We are still pilgrims. We are not there yet. We are seeking a city whose builder and maker is God, and that city is not Rome. *Roma aeterna* is gone forever. It is not Athens, Washington, London, Moscow, Mecca, Baghdad, or Beijing. None of these is that City with Foundations whose builder and maker is God. In the meantime we are called to live by love. Love is the one thing we can experience in time that will remain in eternity. Faith, hope, love, these three; but love is the greatest. Love is eternal.

About a thousand years after Augustine died, Dante presented a compelling vision of this love. In some ways, Dante's vision is a commentary on that medieval depiction of Christ which can be seen, for example, at the cathedral in Salamanca. There sculpted on an early medieval tomb, are the outstretched arms of the cosmic Christ, holding in his nail-scarred hands the moon and the stars. This is the cosmic odyssey. This is how Dante described that vision— tried to describe that vision—in the closing lines of the *Paradisio*:

> Like a geometer who sets himself to square the circle and is unable to think of
> the formula he needs to solve the problem, so was I faced with this new vision.
> I wanted to see how the image would fit the circle, and how it could be that
> that was where it was. That was not a flight for my wings. Except that my mind
> was struck in a flash, in which what it desired came to it. At this point of high
> imagination, all failed; but already my desire and my will are being turned like a
> wheel, all at once, by the love which moves the sun and the other stars.[32]

It is the cosmic Christ who leads us in this cosmic odyssey by the love that was written in bloody garments at Calvary, the love shed abroad in our hearts by the Holy Spirit. It is the same love which moves the sun and the other stars.

Chapter 3

On the Developing Scientific Understanding of Time

RUSSELL STANNARD

All of us share a certain "commonsense" understanding of time. It is divided into past, present, and future. The past consists of all those things that have happened—those things that once existed. These happenings might well have left memories and other lasting effects, but the past events themselves no longer exist. By "the future" we mean all those events that might one day happen, but as yet do not exist. The present is that instant in time we label "now." It divides the past from the future. We live in the present. All that exists is what belongs to the present—what is happening right *now*.

Time moves on. We are being swept towards the future. What lies in the future, becomes the present, and is then left behind in the past. What might be, becomes an actuality, before passing once more out of existence in the past. We call this the "flow of time," or the "passing of time." We might disagree about how fast time flows (it seems to speed up with old age), but we are all agreed that time does indeed pass.

The future is open, uncertain. Through what we do now, we can affect the future. The past, on the other hand, is fixed. We might not agree on what happened then; memories can play tricks, or we might put different interpretations on past events. But whatever it was that happened, we have no power to change it now.

We exist in time; we also exist in space. But space and time are very different from each other. We measure distances in space using a ruler; we measure intervals of time on a watch. There is just the one three-dimensional space, and the one one-dimensional time, common to us all. This means, for instance, that we are all agreed as to the order in which events occur—whether one occurs before the other, or whether they happen simultaneously.

These then are some of our commonly shared ideas about time. In fact, as we shall see, science shows that most of them are *wrong*.

In religious circles, it has been recognized for a long time that God's relation to time could be different. Although we interact with God *in* time—for example, when we engage in prayer, perhaps asking God for some future favor—it is also believed that God is, in a sense, *beyond* time—he is transcendent; he is unchanging. But this raises a difficulty: if he never changes, what is the point of asking him for anything? Our request will not change his mind.

And that is not the only problem over God and time. It is held by Christians that God has foreknowledge—he knows the future. This has never been easy to accept. After all, how can God know the future before I and everyone else have made up our minds as to what we shall do? Surely God could only have such knowledge if humans were nothing more than totally predictable automatons. What then of our supposed free will? For this reason, it is understandable that many people—including those theologians called "process" theologians—reject this claim. And yet God's foreknowledge is part and parcel of orthodox Christian belief. It is what gives believers added confidence in the assurances from God that all in the end will be well for those who love him. It is not a case of God having a better judgment as to what the future might contain; he actually *knows* the future, or so it is claimed.

If we are somehow to accommodate these strange ideas about God in relation to time, then our commonsense notions of time need a shake-up. And that is what modern physics does. Let me explain:

I shall be drawing on Einstein's theory of relativity, but let me reassure you that it is *not* my intention to put you through a crash course on relativity theory! So there is no cause for alarm. It is sufficient for our purposes for me merely to point out to you some of the theory's more startling consequences—consequences which I hasten to assure you have been fully vindicated by experiment.

I want you to imagine an astronaut in a high-speed space craft, and a mission controller in Houston. The astronaut leaves earth at a speed of nine-tenths the speed of light. (The speed of light is 300,000 kilometers per second, so such a rocket engine is clearly not a practical possibility. However, it is more striking to illustrate the effects with somewhat exaggerated examples.) Relativity theory is able to show that, with the astronaut and mission controller in relative motion, they do not agree on the distance the craft has to travel to reach a distant planet. At the particular speed we have chosen (nine-tenths the speed of light), the astronaut calculates the journey distance to be about half that estimated by the mission controller.

Nor is this the only effect of their relative speed. The two not only disagree over their estimates of distance, but also of the time the journey takes. According to the controller, time for the astronaut is passing at about half the rate it does for himself. Thus, everything happening in the space craft—the ticking of the clocks, the astronaut's breathing rate, his heart beats, his aging processes—everything has slowed down by a factor of a half. Not that the astronaut will be aware of this; a slow clock looked at by a brain in which the thinking processes have been slowed down by the same factor, appears perfectly normal.

Which of the two observers is right over their assessments of distance and time? It is impossible to say. But why? Won't the astronaut realize that something must be wrong with his observations when, because of his slowed-down clock, he arrives at his destination in half the time it should have taken? No. Remember, he thinks the journey distance is only half that which the controller claims it to be. Both the astronaut and the mission controller have sets of measurements that are entirely self-consistent.

But, it might be argued, the mission controller is stationary, whereas the astronaut is travelling at high speed; perhaps the speed of the astronaut is making his measurements unreliable in some way—and so one should place greater weight on those of the stationary observer. This is to lose sight of the fact that the mission controller is *not* stationary—he is on the earth's surface, and the earth is spinning; the earth goes in orbit around the sun; the sun goes in orbit around the center of the galaxy; the galaxy is moving in its cluster of galaxies. ... How does one define a "stationary observer"? All uniform motion is relative; the two observers are in relative motion, and that is all we can say—hence the name "relativity theory." There are no grounds for according preference to the measurements of one observer over those of the other.

Confusing? It certainly appears that way when one comes across the ideas of relativity theory for the first time. We are so accustomed to thinking in terms of us all inhabiting one space and one time, that it requires a severe mental wrench to conceive of something radically different: namely, that we each inhabit our own space and our own time, and these will differ from each other if we are in relative motion. Standing still like this, I have the same space and time as you. But once I start to move, my space and time are not yours. The reason most people go through life unaware of this is that for most everyday purposes, the differences between our various estimates of distance and of time are so small as to make no practical difference. For example, those who decide to drive express trains all their working lives, will age less quickly than those who settle instead for sedentary office jobs. But the effect of the train's motion is to increase the driver's life expectancy by only one millionth of a second—hardly a factor worth taking into account when deciding one's career. (The effect is that small because even the speed of an express train is negligible compared to that of light.) As I said before, the effects only become significant at very high speeds—the kinds of speed I have to deal with in my experiments on nuclear particles, where everything travels at speeds close to that of light—nevertheless the effects are there all the time.

So how are we to understand these differing spatial and temporal measurements? Let me give you an analogy:

I am holding up this pencil. What do you see? You all see something slightly different. Some of you see a long shape, others a short one. Confusing? No. We think nothing of such differing observations. We realize that what each person sees is merely a two-dimensional projection of what in reality extends in three dimensions. The two-dimensional projection is at right angles to the line of sight, and this line of sight will vary depending on where the observer is seated relative to the pencil. We are unperturbed by the different observations because we know that when each observer makes due allowance for how the pencil extends along his or her line of sight we all come up with identical results for the true length of the pencil.

Now we can use this as an analogy for the case of the astronaut and the mission controller and their differing perceptions of the distance and the time of the journey. Einstein's solution to the problem is that we are *not* dealing with a three-dimensional space and a separate one-dimensional time. We are dealing with a *four-dimensional reality*: the three dimensions of space plus a fourth dimension closely related to time. We call this combination *space-time*.

What kind of "objects" are we dealing with in four-dimensional space-time? They must depend on the four quantities: the three spatial directions and the time. This means we are dealing with *events*. An event takes place at a particular location, or position in 3-D space; it also takes place at a particular point in time. One event might be the launch of the spacecraft from earth at a specific time. Another would be the arrival at the distant planet at some later time. Having specified these point-like events in the four-dimensional space-time, we can now ask what the four-dimensional "distance" is between the events marking the beginning and end of the journey. We have already seen that the astronaut and the mission controller do not agree on either the spatial separation of these events, or the temporal separation. But this need worry us no longer. Why? Because once you start thinking of reality as being four-dimensional, then the spatial separation between events is merely a three-dimensional *projection* of that reality. Likewise, the temporal separation is only a one-dimensional projection. From our analogy with the pencil, we know that projections are liable to change as one changes one's viewpoint. It turns out that where four-dimensional space-time is concerned, "changing one's viewpoint" means more than simply being at a different spatial position. Because of the way space-time mixes up space and time, "changing one's viewpoint" means having a different speed. Two observers in relative motion (like our astronaut and mission controller) have differing perspectives on space-time by virtue of that relative motion.

So, differing spatial and temporal projections are no matter. What counts is what these observers get when they calculate the actual separation of the events in four dimensions. When they each plug in their own versions as to what the spatial and the temporal separations are, they obtain identically the *same* value for the four-dimensional space-time separation. You recall the astronaut thought the distance covered by the journey was shorter, but he also thought the time of the journey was shorter; the two cancel out and he gets the same overall result as the controller. And what goes for the astronaut and mission controller holds for any one else observing that interplanetary journey. The four-dimensional separation is the one thing they can all agree upon. It is for this reason that one is led to the conclusion that what really counts is not the individual estimates of spatial distance or temporal interval but the four-dimensional space-time separation. Einstein himself once declared that henceforth we must deal with "a four-dimensional existence instead of, hitherto, the evolution of a three-dimensional existence."

It is an extraordinary conception. There is little point in trying to visualize four dimensions—you just get a headache. The best I can do is to hold up four fingers. Reality is like this, rather than like three fingers representing three-dimensional (length, breadth, width going in different directions) space evolving with time—time being something separate. No, four fingers is what we have. But of course this is just a rather an inadequate analogy. I'm cheating. Strictly speaking these four fingers ought all to be mutually at right angles and that is a physical impossibility to demonstrate—it always hurts to try to do impossible contortions. But if you do not pay too much attention to the wrong angles, that is the best one can do by way of visualizing the situation.

Actually, the preferred course of action among professional physicists is to ignore such visual aids, and just allow oneself to be guided by the mathematics. Mathematically speaking one can handle any number of dimensions—you just add on further terms in the equations.

I have already said how one must stop thinking of a separate time outside three-dimensional space. One must also not get lured into thinking of a separate time outside four-dimensional space-time. Time is not out here. That fourth finger is time. What it means is that four-dimensional space-time does not change. Something can only change *in time*. But space-time is not in time. *All* of time is here in this fourth finger. One point on my finger is the moment you walked into this lecture theatre. The knuckle is this particular instant in time. Towards the end of my finger is the moment when you can thankfully head for the doors and go to lunch and take an aspirin after all these mental gymnastics. It's all there: *past, present, and future.* It's static. We call it a *block universe.* Just as each point in space exists on an equal footing with any other, so does each point in time. We are accustomed to think that all of space exists at each point of time. So, for example, at this instant in time not only does Portsmouth exist, but also New York, Hong Kong, planet Jupiter, and distant galaxies. So all of space exists at each point in time. What *this* is saying is that likewise at each point in space, all of time exists. At this point of space here—in this hall—all of time exists: the day work started on building the hall, the instant you entered it this morning, this present moment, your leaving it, the day the hall gets demolished. It all exists here—in some sense. I stress "in some sense." This finger assures us that this is so—but *how?* It defies the imagination; it is all so counterintuitive.

But let us press on regardless—allowing the analogy or the mathematics to be our guide. According to this viewpoint, there is nothing special about this particular instant we choose to label "now." All points in time are on an *equal* footing. The future in itself is *not* uncertain. All right, we might not *know* what the future holds. On the basis of the limited information we have at our disposal at this point in time we are unable to predict what lies in the future. But our modern understanding of time would seem to indicate that whatever the future holds, in some sense it is already there, fixed, waiting for us to come across it.

Now this is so counterintuitive that some people find this conclusion impossible to accept, including certain noted and highly respected physicists, such as John Polkinghorne who will be speaking to you in Cambridge next week. But I think the conclusion is forced upon us.

Why have I gone into this lengthy discussion? Partly because it is a fascinating subject in its own right. But also it has an impact on religious belief. Once it is accepted that in some mysterious sense, it is possible to assert that the future "exists" on the same footing as the present, this must surely go some way towards adding plausibility to the idea that God might have knowledge of it.

But, I hear you muttering, if physics does not pick out any special instant in time to be called "now," where does the concept of "now" come from? And if nothing is changing (the block universe idea), what about the flow of time—where does that come in? Not only that, if the future already exists, where do we get the notion of the future being uncertain and open to being affected by what I do now? What does this do for my free will? Am I not reduced to the level of an automaton?

Here we touch on one of the truly great mysteries: the fact that we have two entirely different approaches to the concept of "time." So far I have spoken of time exclusively in the way the concept is addressed in physics. Let us now see how this same word *time* is used in a different context—the description of what it is to be a thinking human being.

On examining the contents of the conscious mind, we find mental experiences—feelings, decisions, sensory experiences, etc. These occur in sequence. What separates one experience from the next? We call it "time." The experiences occur in time. We are able to estimate and compare these separations or intervals of time. This might be done through noting the extent to which the memory of a past experience has faded—the greater the separation in time, the greater the degree of fading. (Here I am assuming that one is not constantly recalling that experience and keeping it vivid, for then one is more likely now to be recalling the most recent earlier recall rather than the memory of the original experience itself.) So the fading of the memory is one way of estimating how long ago that experience was.

In addition there might be an indication of the time interval based on the number of other notable experiences we have had subsequent to the one in question. If this is the case, then it might give us a clue as to why time seems to speed up as we get older. When one is young and inexperienced, everything is new and catches the attention—one is laying down memories at a furious rate. As one gets older and more worldly, wise, or, dare one say it, more jaded, most experiences are so familiar they just pass us by and probably do not get recorded so faithfully as in the past. So, if you are using a time estimate based on how many notable experiences you have had since the one in question, it will mentally seem that not much has happened since that event, and thus not much time has elapsed, although the calendar shows otherwise.

The precise mechanism by which we subjectively assess time intervals is not at all well understood. All we know is that we do have our own internal, mental way of doing it, semi-quantitatively at least.

For each experience there are other experiences on either side of it along the sequence, with one exception—the experience that marks the end of the sequence. This end point of the sequence we designate "now." It is only in consciousness that the "now" acquires its special status.

Although we use the word *time* to describe the separation between our mental experiences, it does not follow that this is the same "time" as is used in the description of what is going on in the physical world. For one thing, mental states occur in time but not in space. (It would be absurd to ask how much space a big decision like getting married takes up, compared to a small decision such as which tie to wear today.) And yet we know how indissoluble is the link between physical time and physical space.

Only through the recognition that we use the word *time* in two distinct ways can it make sense to say that all of time, including the future, exists now. What this means is that all of *physical* time exists at the instant of *mental* time called "now"—and indeed at every other instant of mental time.

It is perhaps unfortunate that the same word "time" is used in two such dissimilar contexts. The reason it is, of course, has to do with the fact that, despite the distinctiveness of physical and mental time, there is a close correspondence between them. A sensory experience which is part of the mental sequence (for example, the hearing of a shot now) is correlated to a feature of space-time (the firing of a gun at a particular point in space-time). The "now" of mental time is correlated to a particular instant of physical time. Although, as I have said, *all* of physical time exists now, consciousness (not physics) singles out one particular instant as having special significance for the "now" of mental time.

A short while later (according to mental time that is), the "now" correlates to a different physical time. The difference between the two physical times, judged on a clock, when compared with the perceived lapse in mental time, gives rise to a "flow" of time. Note that without two types of time, there could be no flow. A flow is a change of something in a given time. The flow of water from a hosepipe, for instance, is the amount of water emitted in a given time. But what possible meaning could be given to the phrase a *flow of time*? Is it the amount of time passing in a given time? It cannot mean anything—not unless we are talking about *two* different times. When the bored listener to this lecture complains that time is passing slowly, she means that, based on her subjective mental assessment of how much time separates her from the end of the lecture, the physical time registered by the hands of her watch are lagging behind where she reckons they ought to be. There is too little physical time corresponding to the perceived span of mental time. Thus she says time is dragging.

Earlier I said how physical time was seamlessly welded to the three spatial dimensions to form a four-dimensional existence, and that one should not fall into the trap of thinking of this four-dimensional world evolving in time, as though time were still somehow separate. That is true as far as *physical* time is concerned. Four-dimensional reality never changes. And yet, as far as we conscious human beings are concerned, there *is* change. Physical time might be integrally part of that four-dimensional existence, but mental time does appear to stand outside it. It focuses conscious attention and mental experience on one particular instant of physical time, and temporarily correlates it with the mental "now." That conscious focus then appears to move steadily along the time axis in the direction labeled "future." It has been described as being like a searchlight beam scanning along. But it is important to understand that this so-called movement is not out there in the physical world itself; it is a feature only of our conscious experience of that world.

Now you might be wondering what all this has to do with religion and with God in relation to time. I have already mentioned God's foreknowledge and how this appears much more plausible once one recognizes that the future is not uncertain. According to the block universe idea, it is all out there, it is fixed. All God has to do is find a way of looking at it—which I reckon should not be too difficult for God! Just as we can, in our analogy, hold up four fingers as a representation of four-dimensional reality, and we can look at it from outside, so I guess God has the ability somehow to take in the whole four-dimensional reality from some external point of view lying beyond the confines of that four-dimensional world. This is what we mean when we speak of God as being transcendent.

But there is another aspect to this. Religious believers accept that God is a *personal* god. We may have difficulty over the lack of a suitable inclusive pronoun, and are not sure whether to refer to God as a "he" or a "she," but one thing God is not, and that is an "it." In thinking of God, we have in mind attributes such as love, justice, and mercy; and we think of God as having a purpose. None of this makes sense except in personal terms. So, in the same way as we interact with other persons we expect to be able to interact personally with God. These interactions take place *in* time. We expect to be able to speak with God, we are encouraged to ask him for things as a child would ask a parent—in anticipation that if what we request is good for us, we shall receive an answer—a parent who on our asking will influence what later transpires. So God has to be *in* time as well as *beyond* it.

But you might be wondering: How can my asking God for anything *in* time affect the outcome when the future, in *physical* terms, is fixed—set in concrete? As I see it, there is no difficulty in that future outcome already having built into it the effects of your prayer to God now. Indeed, some other Christians and I think nothing of praying for events that have already happened, but about which we do not as yet know the outcome. It might be some oversight that we did not pray about it earlier. We are confident that an all-knowing God will be aware that we will be offering up such a prayer later, and he will take that into account in determining the outcome. Thoughts of a block universe can really transform one's thinking about what can and cannot be done.

But always when I talk on such matters, there are some who find it hard to accept it because it appears to compromise their sense of free will. How can one be free if in some sense the future is known—not known to us, but known to God? Perhaps another analogy will help.

Unbeknown to you, you were videotaped on a security camera arriving here this morning. I have seen the video: you arriving, chatting to each other, chatting to me. I have seen the tape so often I know it by heart. Show me any incident, and I know what happens next. There is nothing you do on those video tapes that I don't know about. There is no element of surprise for me. Does that mean I am looking at a recording of automatons—predetermined robots that have no free will? Obviously not. The tapes are a record of the actions of free people making free-will decisions. The fact that when I play the tape, I know what the outcome of your decisions is going to be, does not in anyway detract from them being your freely made decisions.

In that analogy, I am playing the role of God—the transcendent God. Just as I am outside the confines of the video tape and am able to view any part of it at will, so God lies beyond this four-dimensional world, etched as it is with the record of our lives.

But the analogy is closer. That video tape I was telling you about, shows not only you, but also myself. It is a record of me interacting with you. I don't just watch the video tape; I am also taking part in it. So it is with God. When he surveys the four-dimensional world with its record of our lives, he sees in it himself relating to and interacting with us. If you like, the world is not a cassette he has rented from the video store; it is God's own home video.

Let me now turn to another way in which modern physics affects our thinking about God in relation to time. As most people know by now, the universe began with a Big Bang: all the stuff of the universe began squashed up at a point. There was a mighty explosion. When we look at distant galaxies—or clusters of stars—we find them still rushing apart in the aftermath of that explosion.

But just because I call it an explosion I don't want you to get the impression it was an explosion much like any other—bigger, but essentially the same: an explosion that takes place at a particular location in space. In the case of a bomb, it might be located in the "Left Luggage" office of Waterloo Station. Following the detonation, the pieces of the bomb come flying out so as to fill up the rest of the station. That is a *normal* explosion. But that is not how it was with the Big Bang. Not only was all of matter concentrated initially at a point, but also all of space. There was no surrounding space outside the Big Bang.

Perhaps an analogy will help. Imagine a rubber balloon. Onto its surface we glue some coins. The coins represent the galaxies. Now we blow air into the balloon. It expands. Suppose you were a fly that has alighted on one of the coins; what do you see? You see all the other coins moving away from you—the further the coin, the faster it is receding into the distance. A coin, twice as far away as another, is receding twice as fast. But that is the observed behavior of the galaxies—they too are receding from us in exactly that manner.

So far we have thought of the galaxies as speeding away from us as they move through space. But with the balloon analogy in mind, we now have an alternative way of interpreting that motion. It is not so much a case of the galaxy moving *through* space, as the space between us and it *expanding*. The galaxy is being carried away from us on a tide of expanding space. Just as there is no empty stretch of rubber surface "outside" the region where the coins are to be found (a region into which the coins progressively spread out), so there is no empty three-dimensional space outside where we and the other galaxies are to be found.

Note that with the particular kind of expansion we are considering, where the speed of recession is proportional to distance, the situation is the same from all vantage points. In the case of the balloon, it would not matter which particular coin the fly happens to have alighted upon; all the other coins are receding from it in exactly the same manner. All points on the surface are on the same footing; there is no point on the surface that can rightly be called the "center." In exactly the same way, there is no point out there in three-dimensional space that can be regarded as the center of the universe. There is no privileged location out there where one would be justified in erecting a blue heritage plaque announcing, "The Big Bang occurred here." In truth, the Big Bang happened *everywhere*.

It is this interpretation of the recession of the galaxies that leads us to conclude that empty space is like very thin rubber—rubber that is constantly expanding. At the instant of the Big Bang, all the space we observe today was squashed down to an infinitesimal point. Because of this, it becomes natural to suppose that the Big Bang not only marked the origins of the contents of the universe, it also saw *the coming into existence of space*. Space began as nothing—it had no size—and has continued to grow ever since.

That in itself is a remarkable thought. But an even more extraordinary conclusion is in store for us. Remember this four-fingered approach to space and time. This time dimension is welded onto the spatial ones; they are inseparable. You cannot have space without time; you cannot have time without space. If the Big Bang marked the moment where space was created, so also must time have been created then—that is when the great cosmic clock began ticking. The Big Bang marked the *coming into existence of time*. This in turn means that there was no time before the Big Bang. Indeed, the very phrase "*before* the Big Bang" has no meaning. The word *before* necessarily implies a preexistent time, but where the Big Bang was concerned, there was none.

Now, for those who seek a *cause* of the Big Bang, God, say there is a problem here. Cause is followed by effect. Boy throws stone—cause; followed by window breaks—effect. Note the word *followed*. It refers to a sequence of events in time: first the cause, then the effect. But in the present context we are regarding the Big Bang as the effect. For there to have been a cause of the Big Bang, it would have had to have existed prior to the Big Bang. But this we now think of as an impossibility. There is no time before the Big Bang.

This gets rid of the kind of creator God that most people probably have in mind: a God who at first exists alone. Then at some point in time God decides to create a world. The firework fuse is lit, there is a Big Bang, and we are on our way. God becomes the cause of the Big Bang. But as we have seen, without time before the Big Bang, there could not have been a cause in the usual sense of that word.

So, where have we got to? Have these considerations dispensed with a creator God? Before jumping to that conclusion, let us consider the following quotation:

> "It is idle to look for time before creation, as if time can be found before time. If there were no motion of either a spiritual or corporeal creature by which the future, moving through the present, would succeed the past, there would be no time at all. . . . We should therefore say that time began with creation, rather than that creation began with time."

If the archaic expression "either a spiritual or corporeal creature" had been replaced by a more up-to-date one—such as "a physical object"—one could well have thought that the quote came from some modern cosmologist like Hawking, or from Einstein. In fact, those are the words of St. Augustine. I think you will agree, they beautifully sum up what I have been trying to say. Modern cosmologists find it hard to come to terms with the fact that, where the beginning of time is concerned, it was a theologian who got there before them—1500 years earlier!

How did he do it, bearing in mind that St. Augustine obviously knew nothing about the Big Bang? He argued somewhat along the following lines: How do we know that there is such a thing as time? It's because things change. Physical objects (for instance, the hands of a clock) occupy certain positions at one point in time, and move to other positions at another. If nothing moves (or in the past had *ever* moved), there would be nothing to distinguish one point in time from another. There would be no way of working out what the word *time* was supposed to refer to; it would be a meaningless concept. *A fortiori*, if there were no objects at all, moving or stationary (because they had not been created), clearly there could be no such thing as time.

In this way, Augustine cleverly deduced that time was as much a property of the created world as anything else. And being a feature of that world, it needed to be created along with everything else. Thus it makes no sense to think of a time that existed before time began. In particular it makes no sense to think of a God capable of predating the world.

Yet despite all this, Augustine remained one of the greatest Christian teachers of all time. His realization of the lack of time before creation clearly had no adverse effect on his religious beliefs. To understand why this should be so, we have to draw a distinction between the words *origins* and *creation*. Whereas in normal everyday conversation we might use them interchangeably, in theology they acquire their own distinctive meanings. For example, if one has in mind a question along the lines of "How did the world get started?" that is a question of origins. As such, it is a matter for scientists to decide, their current ideas pointing to the Big Bang description.

The creation question, on the other hand, is quite different. It is not particularly concerned with what happened at the beginning. Rather it is to do with: "Why is there something rather than nothing?" It is as much concerned with the present instant of time as any other. "Why are *we* here? To whom or to what do we owe our existence? What is keeping us in existence?" It is an entirely different matter, one not concerned with the mechanics of the origin of the cosmos, but with the underlying ground of all being.

It is for this reason one finds that whenever theologians talk about God the Creator, they usually couple it with the idea of God the Sustainer. His creativity is not especially invested in that first instant of time; it is to be found distributed throughout all time. We exist not because of some instantaneous action of God that happened long ago—an action that set in motion all the events that have happened subsequently—an inexorable sequence requiring no further attention by God. We do not deal with a God who lights the firework fuse and *retires*. He is involved first hand in *everything* that goes on.

Chapter 4

Time in Physics and Theology

JOHN POLKINGHORNE

Everyone knows that St. Augustine was aware of what time is as long as he did not try to think about it. Much the same can be said of modern scientists. Although the twentieth century brought some important new insights into the nature of time, a number of significant issues remained unresolved. Partly this is because of continuing scientific uncertainties, but it is also partly because ultimately the nature of temporality is a metaphysical issue that cannot be settled by unaided science alone. I shall seek to survey some of these scientific perplexities and the possible meta-scientific responses to them and to consider how these issues relate to religious concerns. As a first step it is necessary to take into account the radical revaluation of our thinking about time that resulted from Einstein's discoveries.

Relativity

Newtonian physics was played out within the container of absolute space and in the course of the unfolding of absolute and universal time. This view is closely allied to commonsense perception and for two centuries there was no serious questioning of the picture it presented. However, difficulties in understanding how to integrate the Newtonian mechanics of particles with Maxwell's equations for the electromagnetic field, led a third-class examiner in the patent office in Berne to make a fundamental reexamination of this point of view. Albert Einstein realized that an operational discussion of how time could be measured was necessary and this resulted in his making a radical reassessment. He supposed that observers were furnished with standard clocks, cyclic mechanisms whose successive revolutions are assumed to measure the elapse of standard intervals of time, but he then raised the question of how differently located observers could ensure that their clocks synchronized with each other. This is clearly an essential requirement if one is to be able to discuss motion taking place through space.

Observer A can send a message to observer B saying that his clock now reads noon, but if B is to use this to set his clock, he must know how long that message took to reach him, so that he is able to make the necessary adjustment. Einstein supposed that the distance between A and B can be measured accurately using standard measuring rods, so B can make the required correction if he knows the speed at which A's message was transmitted. But, in order to measure the speed, one must already have a system of synchronized clocks in operation. The problem has bent round upon itself. Without synchronized clocks one cannot measure speed; without known speed one cannot synchronize the clocks.

Einstein suggested a daring way out of the dilemma. He proposed that in nature there is a universal message carrier whose speed is a known constant of nature that is the same for all observers. This messenger is light. The problem of synchronization is then solved, but at the cost of some highly counterintuitive consequences. If the speed of light is an absolute constant, it will be the same if emitted from a moving source or from a stationary source. This runs counter to commonsense intuition, which supposes that the velocity of the source should be added to the velocity of the emitted light. That seems to be the case in everyday experience when a capsule thrown forward from a moving train travels faster than a capsule thrown from rest. Following all this through led Einstein to predict a number of strange phenomena: moving clocks run slow compared to clocks at rest; moving measuring rods contract along their direction of motion; judgment of the simultaneity of distant events depends upon the state of motion of the observer making the judgment. Newton's universal time, accessible to all, was replaced by the variety of different observers' temporal experiences. All these predictions have been found to be the case. We are only unaware of these strange effects in everyday life because their consequences are very small for processes involving speeds much less than the velocity of light, which is about 300,000 kilometers a second.

These remarkable insights came to Einstein in 1905 (an amazing year in which he also made two other discoveries of fundamental significance). They constitute his *special theory of relativity*. In seeking a name for it Einstein did not know whether to stress its relativity or its absoluteness, for though different observers make different temporal judgments, their accounts are all mutually reconcilable in terms of a four-dimensional space-time measure, called interval, on which all agree and which ensures that all observers take a common view about the order in which causally related events take place.

Special relativity is concerned with the unaccelerated motions of bodies. In 1915, Einstein extended his thinking to take in acceleration, including the effects of gravity. This advance constituted his *general theory of relativity*. The central idea involved what was termed "the principle of equivalence."

Mass enters physics in two ways: as inertial mass, measuring a body's resistance to having its state of motion changed, and as gravitational mass, measuring the strength of the interaction by which gravity affects the body's motion. When you stub your toe against a stone you are encountering its inertial mass, its reluctance to get out of your way. When you pick up the stone you are encountering its gravitational mass as you experience its weight. Although logically distinct from each other, these two kinds of mass are always numerically the same. In a word, they are quantitatively equivalent. As a consequence of this, all bodies move the same way in a gravitational field. For example, a body twice as massive offers twice as much resistance to having its motion changed, but it also experiences twice as strong a gravitational force to bring about that change. The result is, therefore, the same as for the lighter body. Einstein saw that this universal behavior meant that the effect of gravity could be treated as if it were due to a distortion or curvature of space-time itself. The general theory of relativity turns physics into geometry. Most scientists would reckon it to be Einstein's greatest discovery.

The resulting theory ties together space, time, and matter into a single account. Matter curves space-time and the curvature of space-time influences the paths that matter takes through it. Sixteen centuries after Augustine, science confirmed his insight that time and the material creation are intimately linked, so that time came into being with the rest of God's creation. It is no good asking what God was doing before creation, since there (was) no "before" at all, only the timeless reality of God's eternity.

Einstein pressed on to use this linkage to construct an account of the whole universe, thereby deriving from general relativity the first truly scientific formulation of cosmology. Interestingly enough, he got the details wrong! It had seemed clear to him, and to most scientists of the day, that cosmology should describe a static universe, everlastingly the same in its overall characteristics. Physics was to be the last science reluctantly to embrace the reality of history. The previous century and a half had seen geology, and then biology, recognize the evolving character of the world, but the physicists still clung to an Aristotelian concept of an unchanging universe. To get a static solution of this kind out of his equations, Einstein had to tinker with them, making what he later came to think of as "the greatest blunder of my life." Within ten years, Edwin Hubble had discovered the recession of the galaxies, and the expansion of the universe became the foundation for what later came to be called "Big Bang cosmology." Science's conclusion appeared consistent with what Aquinas had believed on the basis of revelation, that time had a beginning. The physicists locate it about fourteen billion years ago.

General relativity's account of cosmology also suggested a qualification of special relativity's observer-dependent concept of now-ness. While this dependence is certainly true for the experience of localized observers, when we consider the universe as a whole there is a natural point of view (or frame of reference, to use more technical language) which defines a natural cosmic time. It is this idea of time to which cosmologists appeal when they say the universe is fourteen billion years old. It arises in this way.

On the largest scale, our universe is homogenous, a fact reflected in the high degree of uniformity found in the cosmic background radiation. The latter is lingering signal left over from the state of affairs in the universe after matter and radiation had decoupled from each other, an event that took place when the cosmos was about half a million years old. This uniformity implies that there is a natural preferred point of view from which to think about cosmology, namely from the perspective of an observer at rest with respect to the background radiation. There is, therefore, a natural global cosmic "now," though this is modified on smaller scales, particularly near entities that grossly distort space-time, such as black holes.

Relativity theory has greatly advanced our physical understanding of space and time, but many puzzles remain. One of them is connected with irreversibility.

IRREVERSIBILITY

With a very small exception that genuinely need not concern us, the basic laws of physics are reversible, that is to say, they do not make a distinction between past and future. If it were possible to make a film of two electrons interacting, that film would make equal scientific sense whether it was run backwards or forwards. Yet our experience of the everyday macroscopic world is certainly irreversible. We know at once that a film in which the bounces of a rubber ball get higher and higher is being run in reverse. In our human experience, there is an arrow of time pointing from past to future. In fact, from a scientific point of view, there are several such arrows, each having a different logical character:

1. *The thermodynamic arrow of increasing entropy (that is to say, increasing disorder) in isolated systems.* This arrow is an expression of the second law of thermodynamics and it is commonly believed, without there being any full understanding of why this should be so, that it underlies our experience of the unidirectionality of time. Entropy increases without external intervention, because there are so many more ways for a complex system to be disorderly than there are for it to be orderly, so that disorder wins hands down. Think of your desk to get the point: unless you intervene it will get more and more untidy.

2. *The arrow of increasing complexity.* The universe started in a very simple state (essentially, an almost uniform ball of energy) and it has now become very differentiated and complicated. Life in general, and human beings in particular, are striking illustrations of this tendency. There is no contradiction here with the second law, for living entities are not isolated systems, but rather they are what is called dissipative systems, dependent for their continuing existence upon their continuous interaction with the environment. We maintain our very considerable order by exporting disorder into our surroundings, for

example when we breathe out carbon dioxide as a waste product of metabolism.

3. *The cosmic arrow of the universe's expansion.* This points foward from the Big Bang.

4. *The arrow of causal ordering, pointing from cause to effect.* While it is our experience that causes precede effects, the fundamental laws of physics do not enforce this to be the case. Their equations admit of two types of solutions: retarded solutions, corresponding to cause before effect, and advanced solutions, corresponding to causes that operate from the future. The empirical nature of our world is fitted by past causes alone, but the reason why this is so is not at all well understood.

5. *The psychological arrow of human temporal experience.* We can remember the past but we have no access to the unknown future.

All five arrows point in the same direction. Yet there is no widely agreed or satisfactory understanding of why this should be so. From a theological point of view, the existence of these arrows seems congenial to the common linear concept of temporality held by the three Abrahamic faiths, Judaism, Christianity, and Islam, which all see time as a path to be trodden. It may seem more problematic from the point of view of cyclical notions of temporality, such as those Eastern faiths appear to espouse when they speak of a samsaric wheel from which to seek release.

Time's arrows can be used to define a sequence of events arranged according to what the Cambridge philosopher, John McTaggart, would have called "the B-series," ordered by the relationship of "before" and "after." McTaggart rejected what he called "the A-series" ordering, based on the division between past, present, and future. We therefore, need to go on to consider questions of "now-ness" and the flow of time.

THE FLOW OF TIME

A number of different scientific points of view have been appealed to by those who consider this issue. In each case a chain of thought is presented leading from science to metascience, and then on to metaphysics more generally, with theological considerations finally coming into play.[1] The links in these chains of argument are not tight connections arising from logical necessity, but they are the looser associations of what one might call alogical consonance. In consequence, there is much room for discussion and dissent. Four broad strategic approaches seem to be possible.

Time as a Psychological Trick

One possibility is that the human experience of the passage of time is a trick of psychological perspective. According to this view, the universe simply *is*, in its four-dimensional space-time character. Special relativity is often appealed to as the scientific support for this metascientific position. The fact that different observers assign simultaneity to different collections of distant events in different ways is held to imply that all these events are existent on equal terms, even if some observers think that they occur at different times and others that they occur at the same time. It is then claimed that this shows that the past and the future must be as physically real as the present.

Another argument leading to a similar conclusion is that the equations of physics contain no reference to a preferred "present moment." So to speak, $t = 0$ has no special significance. Physics does not know about "now."

If these arguments were right, then the proper metaphysical picture would be that of the "block universe": reality as a chunk of frozen cosmic history. The true reality is the whole space-time continuum. This seems to have been the way in which Einstein thought about the matter. When he wrote a consolatory letter to the widow of his great friend Michele Besso, he told her that to physicists the distinction between past, present, and future was an illusion, even if a somewhat stubborn illusion.

There are obvious analogies here with the way in which classical theology, stemming from Augustine and Boethius, conceived of God's relationship to creation. The atemporal deity perceives the whole of history "at once." Aquinas held that this was fully compatible with human freedom because God's knowledge is always contemporary knowledge and therefore no *fore*knowledge of free actions was involved. For God there is no unknown future; all is present before the atemporal gaze of the divine.

Since God surely knows things as they really are, there would seem to be an implication from classical theology that the block universe is indeed the right picture to take. It need not be a deterministic universe that is viewed this way, for determinism is concerned with the causal relationships between events and what these are is not logically settled, one way or the other, simply by adopting the block universe approach. The nature of causality is a separate question, but one has to recognize that there is a certain tendency to associate atemporality and determinism together.

However, I reject the block universe picture, in particular challenging its alleged scientific motivation. If the laws of physics do not contain a representation of the present moment, then so much the worse for physics! Our basic experience of the moving present is not to be dismissed so summarily. There must just be more to say than physics can tell. Only a strong reductionist, for whom physical science is all, could maintain the contrary.

Of course, I agree that it is the case that different observers give different interpretations of simultaneity, but all these interpretations are *retrospective* orderings of what is already causally past. Observers do not know about distant events until those events are unambiguously within their past lightcones (the domain of events that can send signals to that observer). Such constructions can do nothing to establish the claim of the already existent reality of future events.

Time as the Measure of a Closed Universe

A second possibility is that time is the measure of the development of the closed order of the universe. This is the picture of the clockwork cosmos, so chillingly articulated by Laplace's idea of a calculating demon who, given Newton's laws and total knowledge of the present, could then predict the whole future and retrodict the whole past.

According to this view, the future and the past are mere rearrangements of the present, without any independent novelty to them, and so one might again think that the block universe would be the natural metaphysical expression to choose. However, it has been more usual to think of this approach in terms of the unfolding in time of a rigorously deterministic process. Its theological counterpart is, of course, the God of deism, who set creation to spinning and then simply let it continue on in its strictly ordered way.

Isaac Newton himself never took so mechanical a view of reality, but his successors certainly did. It all looks somewhat different today, because twentieth-century science uncovered the existence of extensive intrinsic unpredictabilities present in physical process. These have two sources. One is the notoriously probabilistic character of quantum theory, operating at the subatomic roots of the world. The other, equally surprising to prior expectation, lies in the discovery of chaos theory, operating at the macroscopic levels of classical Newtonian physics and everyday experience. It turns out that, though there are some clocks around, many classical systems are what we might call "clouds." That is to say, they are so exquisitely sensitive to the fine details of their circumstance that the slightest variation or disturbance will totally change their future behavior. It is well known that one way in which this unexpected discovery came to light was through the study of simple models of the earth's weather systems—hence the "butterfly effect": that an African butterfly flapping its wings in the jungle today could create effects that grew and grew until they produced a storm over England in about three weeks' time.

Both quantum theory and chaos theory have been nails in the coffin of a merely mechanical account of nature, if by mechanical we mean systems that are tame and controllable because their behavior is predictable. Unpredictability is what the learned call an epistemological property; that is to say it relates to the knowledge that we can have of a system's behavior. There is no inescapable entailment carrying one over from epistemology to ontology, that is to say what the system is actually like in its intrinsic nature. However, scientists are instinctive realists, so that for them, what we can or cannot know is held to be a reliable guide to what is actually the case. If we did not think that our knowledge of the physical world is telling what it is actually like, it is hard to see why we should go to all the effort and struggle involved in scientific research. So, in a phrase I coined and that I am rather fond of, scientists have the stirring legend "Epistemology models Ontology" emblazoned on their T-shirts. What we can know is a reliable guide to what is the case. Those who are religious believers can appeal to a theological undergirding of this realist belief: the faithful Creator is surely not in the business of misleading us about the nature of creation.

In the case of quantum theory, following the strategy of realism has been an almost universal move, both for physicists and philosophers. Almost all of them regard Heisenberg's uncertainty principle (which was originally concerned with the epistemological question of what could actually be measured) not as merely a principle of ignorance, but as an ontological principle of indeterminacy. That this is not a move forced by physics itself is made clear by the existence of a deterministic interpretation of quantum theory, due to David Bohm.[2] His theory is identical in its empirical consequences to the conventional indeterministic quantum theory, but for Bohm probabilities arise solely from partial ignorance of all that is going on.

In the case of chaos theory, such a strategy of ontological interpretation has been far less popular, though some of us have advocated it.[3] Most people, however, have been bewitched by the Newtonian equations from which the discussion started—but with which it need not end—and they have accorded them a prior status that then forecloses the ontological issue. However, it is worth considering the possibility that these classical equations are no more than what one might call a "downward emergent approximation" to a more subtle and supple physical reality.

The fact of the matter is that questions of causality are metaphysical questions that must receive metaphysical answers, which are then to be defended on metaphysical grounds. Inevitably there is a choice of lines that could be followed. It is possible to regard all unpredictabilities as simply epistemological deficits, no more than unavoidable ignorances. On this basis, those who wish to do so can then retain a picture of the temporal unfolding of a closed deterministic universe. However, it is also possible to take these unpredictabilities as affording welcome ontological opportunities. Those who take this latter course are then led to believe the next possibility.

Time as the Unfolding of an Open Universe

A third possibility is that time is the unfolding development of a universe open to its future. On this view, the intrinsic unpredictabilities of nature are to be interpreted as affording room for the operation of additional causal principles, over and above the exchange of energy between constituents that has been the traditional causal picture of a methodologically reductionist physics. In the case of chaos theory, consideration of the form that these additional causal principles might be expected to take suggests that they will be holistic in character (since they relate to sensitivities that will not permit the system to be treated in isolation from the influence of its environment) and concerned with the generation of future patterns of behavior that may be characterized in terms of an informational input specifying what the pattern will actually be. The intrinsic indeterminacy of a chaotic system means that it faces a portfolio of possible future behaviors, only one of which, of course, will actually be realized. These many different possibilities do not differ from each other in terms of their total energy content but in the character of the patterns in which that energy will actually flow. One might picture the occurrence of one of these patterns as representing an input of information specifying its particular structure.

A convenient way of summarizing these metaphysical expectations is to say that the familiar energetic causality of conventional physics will be complemented by the causal efficacy of an additional causal principle that one might call "active information."[4] One reason for defending metaphysical speculations of this kind is that they offer the *glimmer*—I say no more than that—of the opportunity for accommodating our direct human experience of intentional agency, and our theological intuition of divine providential agency, both operating within the flexible future of an open universe.

A worldview of this kind would have certain implications for theology. It allows the possibility for divine action to take the form of a continuous interaction with creation's history, complementing but not overruling the activities of creatures. It is consistent with a strong concept of continuous creation, and with the theological interpretation of an evolutionary world as being, to use the English clergyman Charles Kingsley's splendid phrase, a creation allowed by the Creator "to make itself,"[5] though not bereft of divine guidance in the doing of it. A number of theological consequences flow from this point of view:

1. Since agency operates within the cloudiness of unpredictable physical process, it can never be exhibited and itemized, as if nature did this, human intention did that and divine providence did the third thing. All are inextricably intertwined, so that faith may perceive, but science cannot demonstrate, their separate actions.

2. The universe is a world of true becoming, in which the form of the future is not already present in the shape of the past. If that is creation's true character, then surely the Creator will know it as it is, and therefore in this unfolding fashion. This seems to imply that God will not just know that events are successive but God will know them in their succession. In other words, there must be a temporal pole within the divine nature in addition to the unchanging eternity of God. Such a dipolar account of divinity seems consistent both with the God of the Bible, continually engaged with unfolding history, and with the divine immersion in time that took place in the incarnation. God's embracing of time in this manner can be understood as part of the Creator's kenotic self-limitation in allowing the created (and temporal) other to be. This idea provides theology with an important resource as it struggles with its greatest perplexity, the problem of theodicy and how we are to understand the evil and suffering present in a world believed to be the creation of a good and powerful Creator. This history of the universe is not the performance of a fixed score written from all

eternity, but it is a grand improvization in which creatures and their Creator all play their parts. We do not have to believe that God specifically wills the act of a murderer, or the incidence of a cancer, but both are permitted to happen in a universe that is allowed to be and to make itself. How to think about the balance between divine action and creaturely activity is the age-old problem of roles of grace and freewill, now written cosmically large.

3. Even more contentiously, this approach raises the question of God's knowledge of the future.[6] God must surely know all that can be known but, in a world of true becoming, the unformed future is not yet there to be known. If this is correct, a further aspect of the Creator's kenotic act of creation is the divine acceptance of a current omniscience (knowing all that can be known now), instead of an absolute omniscience (knowing all that ever will be knowable). Although this is in conflict with the thought of classical theology, it is a view that has a good number of modern supporters.

Speculative Proposals

A variety of highly speculative proposals about the fundamental nature of time and of the universe have arisen from quantum cosmology.[7] The attempt to apply quantum mechanics to the whole universe is not a project of manifest feasibility. Its exploration has been greatly hampered by the fact that, at present, we do not know how to reconcile quantum theory and general relativity satisfactorily with each other. Nevertheless, cosmologists are bold thinkers and there have been a number of speculative proposals. Some see time as a secondary construct, emerging in certain circumstances from a more basic theory based on space alone. Others see universes bubbling up from fluctuations in an ur-vacuum, from which worlds rise and fall in endless profusion. Yet others propose that we live in a multiuniverse, each component of which is a universe with a different unfolding of history according to the way in which the outcomes of quantum events have actually occurred within it.

So far, these ideas are too precarious and uncertain to be the basis for much useful metaphysical or theological thinking. Perhaps one can make two comments from a theological perspective. One is that Christianity has a large stake in the reality of historic process and it would have great difficulty in accommodating the idea of many alternative histories (one in which Judas betrays

Christ and one in which he does not, for instance). The other comment is that it would be unwise to take too constrained and limited a view of the Creator's will for the fertility of creation. However, at present, I believe it is best to adopt a cautious stance and not allow ourselves to be carried away by ingenious but doubtful speculations. We should remember that a great Russian theoretical physicist, Lev Landau, once said about cosmologists that they are "often in error but never in doubt."

Eventual Futility

Scientists do not only peer back into the past but they can also discern the general outlines of the future. As far as the universe is concerned, two alternative scenarios are possible. Which scenario will actually occur depends upon the balance between the effects of the Big Bang, blowing matter apart, and the force of gravity, pulling matter together. These two effects are very evenly balanced. Currently, most cosmologists favor the idea that expansion will be the winner, but it is probably prudent to consider both possibilities. It turns out, however, that, either way, the ultimate cosmic future looks bleak.

If expansion predominates (which is scientifically the currently favored alternative), the galaxies will continue to fly apart forever, eventually decaying into ever-cooling low-grade radiation. That way, the universe ends with a dying whimper. If, however, gravity were to win, the present expansion would one day be halted and reversed. What began with the Big Bang would end in the big crunch of cosmic collapse. That way the universe also ends with a bang. Either way, its ultimate fate is futility.

Neither possibility will happen soon. Many tens of billions of years of cosmic history are expected to lie ahead. But it is as certain as can be that carbon-based life will eventually prove to have been a transient episode in the universe. Contemplating these predictions, the distinguished American theoretical physicist and atheist, Steven Weinberg, said that the more he understood the universe, the more it seemed pointless. From within his horizontal perspective, one can fully understand Weinberg's judgment.

Conclusion

From a theological perspective, however, these gloomy but reliable predictions simply remind us that the fulfillment of the kingdom of God will not come solely from the processes of present history. A kind of evolutionary optimism is not enough. If there is a hope for the universe beyond its death—or a hope for ourselves beyond our deaths—it can only lie in the everlasting faithfulness of God, the One who, as Jesus said, is the God of Abraham, Isaac, and Jacob, the "God not of the dead, but of the living" (Mark 12:27). One of the central points at issue between the atheist and the believer is whether we live in a cosmos or a chaos, whether or not the universe makes total sense, now and always. I believe that it is crucial to Christianity to hold to a well-articulated and credible eschatological hope. The foundations of that hope are the faithfulness of God and the resurrection of our Lord Jesus Christ. But to go on to consider that topic, vital as it is, would be to tell another story, which is something I have tried to do in a recent book.[8] All I want to do now is make a single point about the form of that Christian hope of the life of the new creation. In this world, the world of the old creation, it is intrinsic to our humanity that we are embodied and that we are temporal beings for whom time is a significant and inescapable reality. I think this will also be God's will for us in the life of the world to come. That is why our hope is a resurrection hope, of reembodiment in the transformed "matter" of the new creation, as it is redeemed from the transience and decay that are inevitable properties of the matter of this world. I believe also that our hope is a temporal hope, for we shall live in the time of the new creation. Similarly time will be redeemed. Human fulfillment is not to be a timeless moment of illumination, but it will be the unending exploration of the inexhaustible riches of the divine nature, made available to us in the everlasting unfolding of the redeemed in the history of the new creation. Now that is really something to look forward to.

Chapter 5

God, Time, and Eternity

WILLIAM LANE CRAIG

In the program, the topic that I am listed to be speaking on today is the topic of the elimination of absolute time by the special theory of relativity. However, in the interim I changed my mind about that topic, and having listened to Sir John Polkinghorne yesterday, I was very glad that I did because I think that Professor Polkinghorne very effectively exploded the idea that the special theory of relativity has eliminated Newton's concept of absolute time. As Sir John said, the notion of time, or temporality, is a metascientific or metaphysical notion at bottom and therefore cannot be pronounced upon ultimately by science. Indeed, I would be so bold as to say that relativity theory actually teaches us nothing about the nature of time but everything about our physical measures of time. So I was glad that I had changed my topic from addressing specifically relativity theory to a more general discussion of the topic "God, Time, and Eternity."

God, declares the prophet Isaiah, is "the high and lofty one who inhabits eternity" (Isa 57:15). But being a prophet and not a philosopher, Isaiah did not pause to reflect upon the *nature* of divine eternity. Minimally, to be eternal means to be without beginning and end. To say that God is eternal means minimally that he never came into being and will never go out of being. To exist eternally is to exist permanently.

A Biblical View of God and Time

But having said that, we must note that there are at least two ways in which something could exist eternally. One way would be to exist omnitemporally—that is to say, at every point in time. And if time is extended infinitely into the past and into the future, then a being which existed omnitemporally would exist without beginning and end. He would never come into existence or go out of existence; he would exist permanently. And typically, the Scripture speaks of God in terms of his everlasting, omnitemporal duration. For example, Psalm 90:1-2 says, "Lord, you have been our dwelling place in all generations. Before the mountains were brought forth, or ever you had formed the earth and the world, from everlasting to everlasting, you art God." The picture here in the psalmist's mind is of an omnitemporal God who endures for all time, from eternity past into eternity future.

On the other hand, a being could exist eternally, without beginning and end, if such a being were altogether timeless; that is to say, a being which completely transcended time, which had no temporal location and therefore no temporal extension but just existed outside of time, would have neither beginning nor end. Such a being would simply exist in a single, timeless "present," if you will. Although the Scripture does not speak of God explicitly in terms of such timeless eternity, there are, nevertheless, some biblical passages that do intimate a transcendence of God beyond time. For example, Genesis 1:1 says, "In the beginning God created the heavens and the earth" (AV). And then it goes on to describe his creation of the first day and the second day, and the third, and so forth. Thus this beginning envisioned by the author of Genesis may not simply be a beginning of the material universe, the cosmos, but a beginning of time itself. Now since God did not begin to exist, this would imply that God, in some difficult-to-articulate way, existed beyond the beginning of time—beyond the commencement of time in the universe described in verse 1.

Similarly, in the New Testament there are a number of very interesting passages that speak of God's existence before time. For example, in the doxology at the conclusion of the book of Jude, verse 25, we read, "To the only God, our savior through Jesus Christ our Lord, be glory, majesty, dominion, and authority *before all time* and *now* and *forever.*" In this passage, in an almost inevitable *façon de parler* (manner of speaking), the author speaks of God as existing before all time; in some sense, God exists beyond time. If time is finite and had a commencement, then God, being eternal, must in some way exist beyond time.

So the biblical data are not clear on the nature of divine eternity. There are passages which intimate that God might be omnitemporal and passages which intimate that he might be utterly timeless, and therefore it is impossible to decide this question biblically. We must turn to rational, theological, and philosophical reflection in order to adjudicate the nature of divine eternity.

A THEOLOGICAL/PHILOSOPHICAL VIEW OF GOD AND TIME

Now someone might say at this point, "Why do such a thing? Why not simply rest content with the biblical affirmation that God is without beginning and end and exists permanently, and let it go at that, and not try to decide between these two competing theories of divine eternity?" I want to suggest two reasons why I think it is important that we delve into this topic more deeply and not just rest with the minimalist interpretation.

The first reason is apologetical in nature. Namely, modern naturalism often attacks theism, or belief in God, not simply on the basis of a lack of evidence for the existence of God but because, naturalists sometimes claim, the very concept of God is incoherent and therefore there cannot be such a being that falls under that concept. A good example of this would be the prize-winning physicist, P. C. W. Davies in his book *God and the New Physics,* which was a runaway bestseller when it first came out and catapulted Davies into instant fame as one of the best scientific popularizers of our day. Davies argues that God can be neither temporal nor timeless. He says that God cannot be timeless because God, as described in the Bible, is a person; but persons are inherently temporal in nature. They act and react, they are conscious beings who deliberate and anticipate and remember. They think about things. They intend to do things and then carry out those projects. All of these are temporal activities, and therefore if God is personal, as the Bible claims, God cannot be atemporal, or timeless.

On the other hand, says Davies, neither can God be temporal. For if God exists in time, then he is subject to the laws of relativity theory which govern space and time, and therefore he cannot be omnipotent because he is under the laws of nature. So the theist is confronted with a dilemma. The theist believes that God is both personal and omnipotent; but if he is both of these he can be neither timeless nor temporal, and therefore such a God simply cannot exist. The God of the Bible does not exist.

Now in answer to someone like Professor Davies, it is futile simply to quote Bible verses to him because his argument is that the biblical concept of God is incoherent. So the Christian theologian needs to provide some sort of coherent model, or theory, of divine eternity which will escape Davies's dilemma.

The second reason that we cannot, I think, be silent on this issue of God's relationship to time is a doctrinal reason. That is to say, for better or worse, there have already been a good many careless statements that have been made about the doctrine of divine eternity, so that it is pointless to remain silent now. The cat is already out of the bag! Preachers from the pulpit constantly make statements about "our going to be with the Lord in eternity," and so forth. Many times, I think, these statements are theologically inaccurate. A good illustration of this problem is the book *Disappointment with God* by the popular Christian author Phillip Yancey. Now I want to say immediately that I enjoyed reading *Disappointment with God* and found much of it to be meaningful and poignant. But nevertheless, the centerpiece of Yancey's solution to the problem of disappointment with God—that is to say, disappointment for the gratuitous suffering and evil that God permits in our lives—the centerpiece of his solution is Yancey's doctrine of divine eternity. But when you read his explication of eternity, you find that it is self-contradictory. He actually adopts two analogies for divine eternity that support mutually exclusive views. One of them supports divine timelessness; the other supports divine omnitemporality. So at the very heart of his book is this logical incoherence that leaves the problem of disappointment with God unsolved.

Therefore, as reflective Christians, I think we simply cannot afford, to remain silent upon the nature of divine eternity. We need to engage in the project of sorting out a theory or model of divine eternity which is biblically faithful and logically coherent.

Now, having said that, I want to emphasize that we do not do so dogmatically, because the Scriptures are open on this issue. The theory that we develop will be held tentatively. It will be put forth as a suggested model for the Christian community to scrutinize and assess. And in fact, when you look at the contemporary scene, you find that Christian scholars do differ on their understanding of divine eternity. Traditionally, God's eternity has been understood in terms of timelessness. God simply transcends time; he does not exist in time. He does not exist now, but he exists simply timelessly. Great proponents of this view have been people like St. Augustine, Boethius, Anselm, and Thomas Aquinas as Timothy George has explained in his essay. And on the contemporary scene, such philosophers as Eleonore Stump and Norman Kretzman, Paul Helm, Brian Leftow, and John Yates have all defended the theory of divine timelessness.[1]

On the other hand, there have also been a considerable number of thinkers who have defended divine temporality. Among classical authors we might mention John Duns Scotus or William Ockham. Isaac Newton, the great father of modern physics, in his scholium to the *Principia*, which is printed in your packet of materials for this conference, defended divine temporality. On the contemporary scene, such thinkers as Alan Padgett, Richard Swinburne, Stephen Davis, and Nicholas Wolterstorff have all opted for models of divine temporality.[2]

Now clearly, both of these views cannot be right because they are contradictory to one another. To say that God is timeless is simply to say that he is not temporal. So, one is the negation, or denial, of the other. If God is timeless, he is not temporal; if he is temporal, then by definition he is not timeless. Very often, lay people will say, "Well, why can't God be both? Why can he not be both temporal and atemporal?" Well, the problem with that answer is that unless you can provide a model that makes sense of that claim, it is flatly self-contradictory and therefore cannot be true. It is like saying that something is both black and not black. That is logically impossible, unless you can give some sort of model that would provide a distinction that would make it possible. For example, something might be black on one side and not black on the other side. Or it might be black at one time but later be non-black at another time. So if you are going to maintain that God is both temporal and atemporal, you need to provide some sort of a model that would make sense of that. But obviously, in this case neither of these two alternatives would do because one part of God can not be temporal and the other part atemporal, because as an immaterial being God doesn't have separable parts. He is not made up of parts.

Neither can you say coherently that God is atemporal at one time and temporal at another time because it is flatly self-contradictory to say that he is nontemporal at a certain *time*. That approach is a contradiction in terms. So, both of these views of divine eternity cannot be right. We have to decide whether God is timeless or temporal.

What I would like to do is first to look at arguments for and against divine timelessness and then to look at arguments for and against divine temporality.

ARGUMENTS FOR AND AGAINST DIVINE TIMELESSNESS

Now most of the arguments for divine timelessness that I read in the literature, I find to be either clearly fallacious or at best inconclusive. But there is one argument for divine timelessness that I do find very persuasive, and this is the argument based upon the incompleteness of temporal life. Temporal life is radically incomplete in that we do not yet have our future, and we no longer have our past. Our past is continually receding away, and we are always reaching out toward the future we do not have. Our only hold on existence is the present moment that is ever fleeting, ever vanishing, ever passing away. And yet this is the only hold on existence that we as temporal beings have. Our lives are thus radically evanescent and have a tenuous hold on existence. But this seems incompatible with the life of a most perfect being, such as God is.

This evanescence of temporal life was brought home to me several years ago in an unexpectedly powerful way as I was reading Laura Ingalls Wilder's book *Little House in the Big Woods* to our small children, Charity and John. Now you would not expect this book to be a source of philosophical insight, but as I came to the final, closing paragraphs of this book, I was absolutely stunned by what I read. (It did not have this impact upon my children, but it hit me like a hammer!) This is what she wrote:

> The long winter evenings of firelight and music had come again . . . Pa's
> strong, sweet voice was softly singing:
>> "Shall auld acquaintance be forgot,
>> And never brought to mind?
>> Shall auld acquaintance be forgot,
>> And the days of auld lang syne?
>> And the days of auld lang syne, my friend,
>> And the days of auld lang syne,

> Shall auld acquaintance be forgot,
> And the days of auld lang syne?"

When the fiddle had stopped singing Laura called out softly, "What are days of auld lang syne, Pa?'"

"'They are the days of a long time ago, Laura," Pa said. "Go to sleep, now.'"

But Laura lay awake a little while, listening to Pa's fiddle softly playing and to the lonely sound of the wind in the Big Woods. She looked at Pa sitting on the bench by the Earth, the firelight gleaming on his brown hair and beard and glistening on the honey-brown fiddle. She looked at Ma, gently rocking and knitting.

She thought to herself, "This is now."

She was glad that the cosy house, and Pa and Ma and the firelight and the music, were now. They could not be forgotten, she thought, because now is now. It can never be a long time ago.[3]

What makes this passage so poignant, of course, is that in our "now" the time that Laura Ingalls thought was so real, the time that was "now" for her, *is* long ago. It is gone—gone forever! Ma and Pa are gone. The American frontier is gone. Laura Ingalls Wilder herself is gone. Those years that she called "the happy, golden days" are gone, gone forever, never to be reclaimed. Time has a savage way of gnawing away at existence, making our claim upon existence tenuous and fleeting. And surely this is incompatible with the life of a most perfect being, such as God is. A perfect being must have his life all at once, complete, never passing away or yet to come. In other words, the life of a perfect being must be a timeless existence in which he exists in an eternal now that never passes away.

This argument for divine atemporality strikes me as extremely plausible and powerful. And yet I do not think it is entirely demonstrative because I think that the fleetingness of time is diminished for an omniscient being. Part of the reason that time's tooth seems so savage to us is because we no longer have a complete memory of the past or anticipation of the future in our minds. But for an omniscient being who knows completely past, present, and future as though they were right now, the fleeting nature of time's passage is not so melancholy an affair. God can recall past events and relive them with a vividness and reality as though they were present. Similarly, he foreknows events to come in the future with the same sort of reality with which he can know present events. So for a being who has complete recollection of the past and complete foreknowledge of the future, the passage of time is not so severe and detrimental a defect as it is for us finite, temporal creatures. Nevertheless, in the absence of countermanding arguments for divine temporality, I do think this argument gives some plausible grounds for affirming that God is atemporal.

What objections, then, might be raised against divine timelessness? Well, one of the most popular objections that has been raised in the literature is that timelessness and personhood are incompatible. Persons engage in activities such as anticipation of the future and recollection of the past; in deliberation and discursive thinking; in experiencing conscious feelings. All of these are temporal activities. Therefore, the idea of a timeless person is said to be incoherent.

Well, is this a good argument? I'm not persuaded that it is a good objection. Let us conduct a thought experiment: imagine that God had refrained from creating the world. Imagine God existing without creation. We can think of a possible "world" in which God alone exists, solitary, and without any universe or created order whatsoever. Would God, in such a world, be temporal? Well, if he had a stream of consciousness, clearly he would be temporal because there would be a temporal series of mental events going on in his mind. But let us suppose that God exists changelessly in such a state, that he has a single state of consciousness. Would He, in that case, be temporal? Well, I think that is far from obvious. On the contrary, on a relational view of time in which time is a concomitant of events, such a changeless state would be a state of timelessness. So God existing in such a state would, I think, plausibly be timeless.

Someone might say, "A personal being cannot exist in a timeless way." Well, why not? What are the conditions sufficient for personhood? Well, it seems to me that the condition which is necessary and sufficient for personhood is self-consciousness. To know oneself as a self, to have self-awareness and self-consciousness and, hence, intentionality and freedom of the will is sufficient for personhood. But self-consciousness is not an inherently temporal notion. God can simply know all truth in a single intuition of truth without having to learn it or having to come by it through a process. As long as his consciousness does not change, there is no reason to ascribe to God temporality. So there is nothing about a self-conscious life that entails temporality as long as it is a changeless self-consciousness.

As for the other properties we mentioned, I would say that while these are *common* properties of human persons (who are, after all, temporal), these are not *essential* properties of personhood. For example, take deliberation and discursive thinking; this is excluded from God not so much because of his timelessness but because of his omniscience. An omniscient being does not need to deliberate because he already knows the conclusions to anything that he might think about. And therefore God's thought life cannot be discursive if he's an omniscient being. He simply knows all truth in a single intuition at a single moment. Similarly, memory and anticipation are not essential to a time-less person because he has nothing to forget and nothing to anticipate if he simply exists timelessly. There is no past and future. So these qualities, though common to human persons, are not essential to personhood, and therefore it seems to me that there is no incoherence in speaking of God as a timeless, personal being.

In fact, I think that the doctrine of the Trinity can help us out here, because the doctrine of the Trinity provides a useful model for God's timeless existence. Very often, people will say that persons have to exist in interpersonal relation-ships, and therefore God would have to be temporal. But what that assumes is that the persons to whom God is related would have to be human persons. But according to the Christian doctrine of the Trinity, that is not true. God, in his own being, is tri-personal, and in the unity of his own being God can enjoy the fullness of interpersonal relationships within the Godhead itself in a timeless and changeless way. Everything the Father knows, the Son and the Spirit know; what the Father loves, the Spirit and the Son love; what the Son wills, the Father and the Spirit will. This is the doctrine of *perichoresis*, accord-ing to which the three persons of the Godhead are completely transparent to one another and interpenetrate one another. And just as we sometimes speak metaphorically of two lovers who sit just staring into each other's eyes, not speaking a word, as "lost in that timeless moment," so, in a literal way, God in the interpersonal relationships of the Trinity, can exist in a timeless moment of complete love, fulfillment, and blissfulness in the self-sufficiency of his own being. Thus I am not at all persuaded that timelessness and personhood are incompatible; it seems to me quite possible, and plausible, that God can exist timelessly while being personal.

So in summary, then, we have seen one good argument for divine timeless-ness—not decisive, but, I think, a plausible argument—and so far no good reason to reject divine timelessness.

ARGUMENTS FOR AND AGAINST DIVINE TEMPORALITY

What about divine temporality? Let me share with you two arguments in favor of divine temporality. The first argument is the argument based upon God's causal relationship to the world. In order to understand this, you need first to understand the difference between *intrinsic* and *extrinsic* change. Something changes intrinsically if one of its properties changes, which it has in isolation from its relationship to anything else. For example, a ripening apple turns from green to red; that is an intrinsic change in the apple. Something changes extrinsically if it changes in its relations to something else. For example, I was once taller than my son John, but I am now shorter than my son John, not because of any intrinsic change in me, but because of an intrinsic change in him. He has grown taller. I have become shorter than John by undergoing an extrinsic change. I have remained intrinsically changeless in terms of my height, but I have undergone extrinsic change in relation to John in that, because of his change in height, I am now in a new relation, namely *smaller than*, whereas before I stood in a different relation, *taller than*, to my son. So thus I have undergone a relational or extrinsic change.

Now in order for something to be temporal, it does not need to be intrinsically changing. All it needs to experience is extrinsic change in its relations. For example, imagine a rock existing in outer space, frozen at absolute zero. (Now I know that is physically impossible, but this is just a thought experiment.) Let us imagine this rock is frozen at absolute zero, so it is absolutely changeless intrinsically. Would that rock be timeless? Well, I think clearly not, because it could still change extrinsically in its relation to things around it. A meteor whizzes by—a little later, another meteor whizzes by—and a little later, another meteor whizzes by. Even though the rock is intrinsically changeless, it clearly stands in temporal relations with these successive events. And therefore merely extrinsic change is sufficient for a temporal existence.

Now God, as the creator of the universe, is causally related to the world. He brings the world into existence. And the question is, would God be temporal in virtue of his changing relationships with a temporal universe? Let us do another thought experiment.

Imagine God existing once more, alone, without the world, without the creation. Now in such a state, God is either timeless or temporal. If he's temporal, then the issue is decided. God is in time. So let us suppose that he's timeless. And now let us suppose that God decides to create the world, and he brings the universe into being. Now when he does so, God either remains timeless or else he becomes temporal in virtue of his new relationship to a changing world. If God becomes temporal, then clearly he is in time. So could God remain timeless while creating the universe? Well, I do not think so. Why? Because in creating the universe God undergoes at least an extrinsic change—a relational change. At the moment of creation he comes into a new relation in which he did not stand before because there was no "before." It is the first moment of time. And at the first moment of time, he comes into this new relation of *sustaining the universe* or at least of *coexisting with the universe*, a relation in which he did not stand before. And thus, in virtue of this extrinsic, relational change, God would be brought into time at the moment of creation.

Thinkers such as Thomas Aquinas attempted to elude the force of this argument by denying that God sustains any real relations with the created order. Aquinas granted that if God does come into new relations at the moment of creation, like *being Lord*, then he would be temporal. So Aquinas was driven to deny that God sustains any real relations to the world. Aquinas said that we as creatures are related really to God as his effects, but God is not really related to us as our cause or Creator. But I think that such a doctrine is clearly an expedient of desperation. God is causally related to the universe, and it seems impossible or incoherent to say that there could be real effects without a real cause. How could we be really related to God as effect to cause, but God not related to us as cause to effect? Moreover, God seems clearly related to us in that he knows us, he loves us, and he wills our existence. So it seems to me that Aquinas' solution is simply not plausible. These are real relations by any sensible definition of the term *real relation*. Therefore I think we have a powerful reason for thinking that in virtue of his causal relationship to a temporal creation, God is temporal.

The second argument that I'd like to share is the argument based upon God's knowledge of tensed facts. In order to understand this argument, we need to appreciate the difference between "tensed facts" and "tenseless facts." For example, it is a tenseless fact that the C. S. Lewis Summer Institute in Cambridge begins on 21 July 2002. That fact never changes. It has always been true, it will always be true, it is tenselessly true that the C. S. Lewis Summer Institute in Cambridge begins 21 July 2002. But that tenseless fact is not enough information to prompt me to leave Atlanta, board a plane on the twentieth of July, and fly to Cambridge. Why not? Well, because that tenseless fact is *always* true. What do I need to know, in addition to that tenseless fact, in order to prompt me to board the plane to fly to Cambridge? What I need to know is the tensed fact that today is July 20, or tomorrow is July 21. In virtue of knowing that tensed fact, I board the plane and come to Cambridge for the conference. So, tensed facts are facts about the relationship of certain events to the present moment. In the English language tensed facts can be expressed by verbal tenses, like the past tense, the present tense, or the future tense; or by adverbs like *today*, *yesterday*, and *tomorrow*; or by prepositional phrases like *in two days' time*, or *three days ago*. All of these are ways of expressing tensed facts.

Now notice that I, in virtue of knowing tensed facts, must have a temporal location. If I know today is July 20, then I am located at July 20. Moreover, in knowing tensed facts, I would be constantly changing. I would know that today is July 20. The next day I would then know that *today* is July 21 and the next day that *today* is July 22. So any being that knows tensed facts is undergoing change and is therefore temporal. As an omniscient being, God cannot be ignorant of tensed facts. He must know not only the tenseless facts about the universe, but he must also know tensed facts about the world. Otherwise, God would be literally ignorant of what is going on *now* in the universe. An atemporal God would not have any idea of what is *now* happening in the universe because that is a tensed fact. He would be like a movie director who has a knowledge of a movie film lying in the canister; he knows what picture is on every frame of the film lying in the can, but he has no idea of which frame is *now* being projected on the screen in the theater downtown. Similarly, God would be ignorant of what is *now* happening in the universe. That is surely incompatible with a robust doctrine of divine omniscience. Therefore I am persuaded that if God is omniscient, he must know tensed facts and, therefore, must be in time.

So we have two good arguments, I think, for divine temporality. What objections might be raised against God's being in time? Again, let me mention two. The first objection to God's being in time is that both of the two arguments that I just gave for divine temporality presuppose a dynamic view of time. As we have already heard in the course of this conference, philosophers of time differ with respect to two radically distinct approaches to the nature of time. According to a dynamic theory of time, temporal becoming is objective and real. The past no longer exists; the future does not yet exist and is pure potentiality; and things come into being in the present and go out of being as they lapse away, so that temporal process is dynamic and real. Past, present, and future are objective features of reality. John Polkinghorne and Bob Russell enunciated this view.

By contrast, theorists who hold to a static view of time think of all moments in time as equally real, whether past, present, or future. Time is like a spatial continuum, and events are ordered by *earlier than* and *later than* on this continuum; but the distinction between past, present, and future is just a subjective illusion of human consciousness. In reality, the universe is a four-dimensional block that just exists. It never comes into being and never goes out of being. It is really coeternal with God, and it can be said to be created only in the sense that it eternally depends upon God for its existence. It has a beginning only in the sense that a meter stick has a beginning, namely, there's a first centimeter. But it does not come into existence; the four-dimensional space-time block just exists. Similarly, on the static theory of time there really are no tensed facts. Linguistic tense serves only to express the subjective perspective of the user. There is no objective truth about what is now happening in the universe, for "now," like "here," serves merely to pick out the subjective perspective of some person. Every person at every time in the space-time universe regards his time as "now" and others as "past" or "future." But in objective reality there is no "now" in the world. Everything just exists tenselessly. Russell Stannard enunciated this view.

If one adopts a static view of time and so denies the objective reality of temporal becoming and tensed facts, then the two arguments for divine temporality are undercut. The argument based on God's real relation to the world assumed the objective reality of temporal becoming, and the argument based on God's knowledge of the temporal world assumed the objective reality of tensed facts. But if a static view of time is right, nothing to which God is related ever comes into or passes out of being, and all facts tenselessly exist, so that God undergoes neither extrinsic nor intrinsic change. He can be the immutable, omniscient Sustainer and Knower of all things and, hence, exist timelessly. If time exists as a four-dimensional block, God does not change in his causal relations to the world. Existing outside of time, he just causes everything to occur in the four-dimensional block at its various space-time locations. But he is absolutely changeless in his causal relations to the world. Similarly, on the static view of time there are no tensed facts. Tensed facts are a subjective illusion of human consciousness. There really is no "now" in the space-time block. There is no past and future. Those are just perspectives of different people in the block, but none of them is objective and real. So if you adopt a static view of time, the arguments I presented for divine temporality are undercut.

Therefore, I am persuaded that one's theory of divine eternity will stand or fall with respect to the decision one takes with regard to a dynamic versus a static theory of time. If you adopt the dynamic theory of time, you should believe in divine temporality. If you adopt a static theory of time, then the most plausible view would be divine atemporality.

Now in my talk this morning I do not have time to delve into this issue. This would take a whole lecture, a whole seminar itself. But if you are interested, I go into the arguments for and against a static and dynamic theory of time in my book *Time and Eternity*.[4] And, for what it is worth, my judgment is that the arguments for a dynamic theory of time are superior to the arguments for a static theory of time. I think that time is dynamic, that the static theory of time is open to severe philosophical objections and, I even think, theological objections, whereas the dynamic theory of time comports with both our experience as well as with what philosophy tells us about the nature of time. Therefore I am persuaded that time is dynamic, and, hence, I come down on the side of divine temporality.

But there is a second objection to divine temporality that we need to deal with before we can conclude, and that is the question: why did God not create the world sooner? The German philosopher Leibniz pressed this objection against the Newtonian philosopher Samuel Clarke in their correspondence. Clarke, like Newton, believed that God had endured through an infinite, empty, dead time up until a certain moment, at which he created the universe. And Leibniz said, "Why didn't he create the world sooner?" Why would God endure this period of creative idleness for infinity before he created the world, and why would he create the world when he did, rather than sooner or later? Look at it this way. On this Newtonian view, at any time t prior to the moment of creation, God delayed creating until some later moment $t + n$. At any moment in the infinite past you pick, God at that moment could have created the world but nevertheless chose not to. Though God has willed from eternity to create a universe, he deliberately refrained from creating at that moment and delayed until some later time. But surely God must have had a good reason for doing something like that. A supremely rational being, such as God is, would not delay carrying out his will for no good reason. But in an infinite, empty time, there can be no reason for preferring one moment rather than another at which to create, for in an infinite, empty time, all moments are alike. They are indistinguishable, and thus there can be no reason for preferring one moment rather than another, and thus no reason for God to delay creating at some time t until $t + n$. Therefore, Leibniz argued, you must say that time began at the moment of creation, that God has not endured through an infinite, empty time up until creation, but rather that time began at the very moment of creation. This is exactly the view that St. Augustine also adopted in dealing with this problem.

But now we have an extremely bizarre situation. We have seen that time must have had a beginning. God exists in time. And yet God is beginningless. How do you make sense of that? How can God exist in time, time have a beginning, and yet God be beginningless? It does not seem to make sense. Does this force us to say that therefore God is simply atemporal?

A Model for Divine Eternity

Well, I think not, and I want to propose a model for divine eternity that I think will resolve this problem. Let us suppose that time begins at the moment of creation, and let us call that moment "the Big Bang" for the sake of convenience. Then God would not exist literally before the Big Bang, because to exist *before* the Big Bang is to be in a temporal relation. So God would not be temporally before the Big Bang. He would in some mysterious way exist *beyond* the Big Bang, but not *before* the Big Bang. Now in such a state, he would clearly have to exist in a changeless way, because if there were events, if he were changing, then time would not begin at the Big Bang. It would begin with those first events. So God existing beyond the Big Bang must exist changelessly. But such a changeless, eventless state is, as I say, plausibly taken to be a state of timelessness. Therefore the model I want to propose is that *God exists timelessly without creation and temporally subsequent to creation.*

I think we can get a physical analogy for this from the notion of an initial cosmological singularity. The cosmological singularity in which our universe began is, strictly speaking, not part of space and time, and therefore it is not earlier than the universe; rather, it is the boundary of space and time. The singularity is *causally prior* to our universe, but it is not *chronologically prior* to the universe. It exists on the boundary of space-time. Analogously, I want to suggest that we think of eternity, like the singularity, as the boundary of time. God is causally prior, but not chronologically prior, to the universe. His changeless, timeless, eternal state is the boundary of time, at which he exists without the universe, and at the moment of creation God enters into time in virtue of his real relation to the created order and his knowledge of tensed facts, so that God is timeless without creation and temporal subsequent to creation.

Now this remarkable conclusion, I think, deserves serious reflection. It means that God, in creation as in the incarnation, has undertaken an act of condescension for our sake. Existing alone in the fullness of the inter-Trinitarian love relationships, God has no need of temporal persons to relate to. In his perfect timeless existence there is no deficit in his mode of existence—no deficiency to be filled. But out of his love and grace he chose to create a temporal world of finite creatures so that they might be invited to share the inner Trinitarian life of the Godhead and the love of the three persons of the Trinity. So God, in creation, stoops to enter into, and to undertake, our temporal mode of existence in order to relate to us and bring us into relationship with himself. And, of course, in the incarnation he stoops even lower still to take on, not merely our mode of existence, but our very human nature itself.

This, I think, makes good sense of the relationship of God and time. God is timeless without creation and temporal subsequent to creation. Having entered into time, he is not dependent upon finite velocity light signals or clock synchronization procedures for knowing what time it is. Rather, existing in absolute time, God is, as Newton proclaimed, the Lord God of dominion of his universe. In the words of St. Jude: "To the only God our Savior through Jesus Christ our Lord, be glory, majesty, dominion, power and authority, before all time and now and forever" (Jude 25).

DISCUSSION

Q: (Hugh Ross): Bill, when the Bible speaks about time, is it not possible that it's restricting itself to cosmic time? And since we can conceive of time as multidirectional, multidimensional, and stoppable, properties which cosmic time does not possess, can we not conceive of temporality independent of cosmic time, and timelessness, therefore, could be simply existence beyond cosmic time? Doesn't Scripture speak temporally of "before the beginning of time"? Aren't these possibilities, at least?

Craig: Yes. Certainly it's possible to think of God as existing in some sort of a second time dimension that would be a sort of hyper-time in which our ordinary time is embedded. But I'm not persuaded, as you know, Hugh, that this is a good alternative, a plausible alternative. I think that it's metaphysically extravagant to postulate a hyper-time, a second dimension of time. There's no scientific evidence for it. In multidimensional string theories, as you know, these additional dimensions are *spatial* dimensions, not *temporal* dimensions. They all evolve in the single time dimension that begins with the Big Bang. So it's a metaphysical extravagance to postulate a hyper-time.

Secondly, I don't think positing a second time dimension really solves anything because all of the problems we talked about regarding the first dimension of time will simply recur with regard to the second dimension of time. Is the hyper-time a tenseless or a tensed time? Is it dynamic or static? And the whole thing just recurs over again. So I don't think it really solves anything.

Finally, my third point would be that I do think positing a second time dimension is open to certain objections, namely, I think that you can only make sense of a hyper-time by construing the unidimensional time in which we live and exist as a static time. If our time is a dynamic time, then it can't be embedded in a higher time dimension. To think of it as a higher time dimension is treating it like a spatial dimension, in which you can take, say, length and add to it width, so that you get a plane. But time isn't stretched out like a linear figure spatially if you have a dynamic theory of time. It will only work on a static theory. And since I don't think the static theory is correct, for numerous reasons, therefore I don't think, ultimately, that hyper-time is metaphysically possible. So for those reasons I would reject it.

Q: (Hugh Ross): Well, how about the possiblity of a hyper-hyper time? In other words some God capacity completely independent of any concept of time we have but which would nevertheless allow God to—

Craig: See, when you use the idea of extradimensionality, I think you're really using it as a metaphor for something that's not literally a higher time dimension. It's a metaphor for saying that God has the capacity to work in our time in ways that are extraordinary, or something of that sort. And certainly I would grant that, but I don't think that the metaphor of embedding higher time dimensions is a useful metaphor because it's too misleading. If taken literally, as I say, I think it's extravagant, it doesn't solve the problem, and it has serious objections.

Q: Thank you for your excellent talk! The concept of the immutability, the changelessness of God, is, I think, essential if we are to stay away from process theology or other areas where I think we could go wrong. If God is timeless before creation and temporal after creation, are you implying some change in his nature, his essence, or his character, or simply his relation to time?

Craig: Very good question! I am not in any way implying a change in God's nature. Remember, I spoke of his undergoing *extrinsic* change, change in relationships. This wouldn't be a change in his nature. I do think God also changes in intrinsic ways—for example, knowing what time it is. He knows it's now t_1, now it's t_2, now it's t_3. But I think that these kinds of trivial changes are not at all threatening to an orthodox concept of God. What is crucial is that

God not change in his attributes of omnipresence, omnipotence, holiness, love, eternality, necessity, and all the rest. Those would all be preserved as essential attributes of God on this model.

Q: You said at the beginning of your lecture that laymen quite often ask the question, "Why could God not be both timeless and in time?" I would ask it again: Why could God not be timelessly existent or temporal? I think there is an element of timelessness within time. That's my argument.

Craig: Well, did you notice that the model I adopted in the end really is that layman's intuition? Namely, I argued that God is *both* atemporal and temporal. That is the layman's intuition, but unless you give a model, that's just a flat contradiction. That's like saying that something is A and not-A, and that's logically incoherent. That's impossible. But I've tried to provide a model so that it's no longer self-contradictory. How do I qualify it? God is timeless without the universe and temporal subsequent to the beginning of the universe. What I've done, in a sense—and I think this is so ironic because I didn't set out to do this—is wind up vindicating what the layman thinks when he says God is both temporal and atemporal. I think that's right; he's atemporal without creation and temporal subsequent to the moment of creation.

Q: But what I would say to you is, something must be lost then, in that transition from being timeless to being temporal, because if God becomes temporal after creation or during creation, then he must no longer remember the timelessness that he had before. He can't remember it because he's no longer timeless.

Craig: Yes—well, that's right! This is a very odd theory, I admit. This is a very, very odd model. But when you're dealing with subjects like time and eternity, almost everything that you come up with is odd! So what this model would require us to say is that God's omniscience in his timeless state would involve knowledge of exclusively tenseless truths, like "At $t=0$ I *create* the world," "At $t=n$ I *release* the children of Israel from bondage," "At $t=n+m$, I *become* incarnate in the person of Jesus of Nazareth," and so forth. At the moment of creation, all of a sudden there would be a vast number of tensed propositions which would switch their truth value from being false to being true: namely, "I *shall release* the children of Israel," "I *shall become* incarnate," and so forth. Past tense propositions will become true: "I *did create* the world one minute ago," "I *did do* this or that," and so on. But there wouldn't be any past tense propositions about this timeless state before the world because it isn't in the past.

Q: What about future contingent propositions? Is God surprised at what we do?

Craig: No, I don't think so because I think he is omniscient. The doctrine of omniscience says that for any true proposition or fact, God knows that proposition or knows that fact, and he does not believe any false proposition. That's the traditional definition of omniscience. Since there are now truths about the future, God, as an omniscient being, must know them. And this is what the Bible affirms. The New Testament has a whole vocabulary of Greek words with the prefix *pro-* like *prognosis*, which literally means "foreknowledge," and it ascribes this to God. He foretells (*promartureo*) the future. He foreordains (*proorizo*) the future. Moreover, God's knowledge of the future is illustrated in prophecy, such as Jesus' prediction of the denial of Judas and the betrayal of Peter, highly contingent events. So I affirm that God is not surprised by what transpires in the course of the unfolding of time because he is omniscient.

Q: Where does he get this foreknowledge once he has become temporal?

Craig: That's a good question. There are at least two theories, I think, that you could adopt for the basis of divine foreknowledge. One would be that God simply has omniscience as an essential attribute; it is an essential attribute of God to believe only and all true propositions. He doesn't learn anything because he just has the essential property of knowing all truth, and it would be wrong to think that God has to somehow learn what he knows. The other model is called "middle knowledge," which holds that God knows what every free creature would freely do in any circumstances God might place him in. In virtue of knowing those truths and of knowing the decree of his own will to create certain circumstances and place certain creatures in them, God then knows everything that will happen. I'm persuaded that either of those two models is a viable model for divine omniscience and the middle knowledge model is especially useful in explaining God's providence over a world of free creatures.

Editor's note: Craig has written extensively on middle knowledge and is its most prominent modern proponent.

Chapter 6

Eschatology and Scientific Cosmology: From Conflict to Interaction

ROBERT JOHN RUSSELL

Blessed be the God and Father of our Lord Jesus Christ! By his great mercy he has given us a new birth into a living hope through the resurrection of Jesus Christ from the dead.

—*1 Peter 1:3*

INTRODUCTION

The defining kerygma of Christian theology is the resurrection of Jesus of Nazareth.[1] For many biblical scholars and systematic theologians, the resurrection is understood as "bodily": the transformation of the total person of Jesus into a new form of existence, eternal life with God. The resurrection of Jesus, properly understood, is more than the "resuscitation" accounts of the raising of Lazareth or the daughter of Jairus. Instead it is bodily: it was experienced by the disciples as recorded in the Easter appearances and delimited from all

95

"spiritualized" or, more reductively, all "subjectivized" interpretations by an insistence on the Empty Tomb traditions. The resurrection is also more than a "miracle" confined to the person of Jesus and played out against the backdrop of a totally ordinary surrounding world. Instead the accounts of the risen Lord include the disciples and their surroundings in what appears to be an environment already showing signs of radical transformation as suggested by the accounts of his appearances and ascension. This view of the resurrection, for those biblical scholars and systematic theologians who pursue it, leads to an eschatology in which all of creation is to be transformed into the new creation, the "environment" called the "new heaven and the new earth." This new creation was instantiated by God's proleptic act at Easter, signaling what will become the future for all creatures when Christ "comes again in glory."

Ironically, for all the hard-won progress in the constructive engagements termed *theology and science* over the past half century, the challenge raised by science in general and scientific cosmology in particular to bodily resurrection and new creation eschatology has received, with only a few exceptions, strikingly little sustained attention. Throughout the theology and science literature, the concept of "creation" refers unequivocally to the expanding Big Bang universe some thirteen billion light years in size (or, if one follows quantum cosmology, the "Universe" or "megauniverse" of endless inflation) and not, certainly, just to planet Earth. Hence, in this literature at least, the creation which will be transformed into the new creation must unequivocally refer to that same universe (or "Universe"). But this claim runs directly up against scientific cosmology for, regardless of the model chosen (again Big Bang, inflationary Big Bang, quantum cosmology, etc.), the future of the universe is certainly not anything like the biblical/theological "new creation." Indeed at least one very prominent scholar in theology and science has acknowledged that "should the final future as forecasted by the combination of big bang cosmology and the second law of thermodynamics come to pass . . . we would have proof that our faith has been in vain. It would turn out to be that there is no God, at least not the God in whom followers of Jesus have put their faith."[2]

The purpose of this paper is to take some additional initial steps in the process of constructing such a response, following on previous work where I have begun to sketch some of these steps.[3] After a brief overview of Big Bang cosmology (I), I will report a bit more fully on the positions taken in New Testament research regarding the bodily resurrection of Jesus (II). My intention is not to enter into the New Testament debates as a New Testament scholar, which I certainly am not. Instead it is to adopt the "worst case" scenario, the one which makes Christianity the most vulnerable to its atheistic critics, namely the bodily resurrection of Jesus and its implications for eschatology, since this position, as I have already claimed, raises serious, and perhaps unsolvable, conflicts and contradictions with science. It is this position—and not those already "jury-rigged" to fit with science—that is worth pursing, since it represents a "test case" of the highest order for those of us who urge that theology and science should be in a posture of "creative mutual interaction" and not in one of "conflict" or "irrelevance."

I will then turn to theology and science (III) to suggest why such an engagement is actually unavoidable, and to report on the work of one scholar, John Polkinghorne, whose approach seems to me to be uniquely promising. I will then suggest that an expansion of the current methodology in theology and science is required, and out of this I will develop a series of guidelines for reconstructing eschatology in light of science and for exploring new approaches to science and cosmology in light of eschatology (IV). These guidelines lead to some suggestions for future research programs in theology and in science (V).

The Challenge of Big Bang Cosmology

"(I)f it were shown that the universe is indeed headed for an all-enveloping death,
then this might . . . falsify Christian faith and abolish Christian hope."[4]
—*John Macquarrie*

To consider the universe from a scientific perspective, we must turn to physics with its theory of gravity and thus Big Bang cosmology.[5]

By 1915, Albert Einstein had constructed his General Theory of Relativity (GR) using his Special Theory of Relativity (SR) as its basis and applying it to the problem of gravity. Here he extended Newton's principle of the relativity of motion to include not only electromagnetism (as in SR) but now gravitation as well. Instead of matter moving within a three-dimensional absolute, container-like space with time flowing uniformly and independently, Einstein's radically new approach combines space and time into space-time[6] and sets it on an equally ontological footing as matter. As a famous interpretation goes, in general relativity "space tells matter how to move and matter tells space how to curve."[7] During the 1920s, telescopic observations by Edwin Hubble showed that galaxies surrounding our Milky Way were receding from us at a velocity v proportional to their distance D, as given by Hubble's law: v = H x D. The expansion of the universe, as described by Einstein's general theory of relativity, had been observationally discovered!

There are three possible types of expansion: 1) Closed model. In this model the universe has the shape of a three-dimensional sphere of finite size. It expands up to a maximum size, approximately 100–500 billion years from now, then recontracts; 2) and 3) Open models. In both the "flat" and "saddle-shaped" models, the universe is infinite in size and it will expand forever. All three came to be called "Big Bang" models because they describe the universe as having a finite past life of 10–15 billion years and beginning at time "t = 0" in an event of infinite temperature and density, and zero volume.

Since the 1970s, a variety of problems in the standard Big Bang model have led scientists to pursue "inflationary Big Bang" and "quantum cosmology."[8] According to inflation, at extremely early times (roughly the Planck time, 10^{-43} seconds after t = 0) the universe expanded exponentially, then quickly settled down to the slower expansion rates of the standard Big Bang model. During inflation, countless domains may arise, separating the overall universe into many universes, each huge portions of space-time in which the natural constants and even the specific laws of physics can vary. In most quantum cosmologies, our universe is just a part of an eternally expanding, infinitely complex megauniverse. Quantum cosmology, however, is a highly speculative field. Theories involving quantum gravity, which underlie quantum cosmology, are notoriously hard to test empirically.

Even if inflationary or quantum cosmologies prove of lasting importance, the far future of the visible universe in which we live is described by Big Bang cosmology. There are two scenarios for the far future: "freeze" or "fry." If the universe is open or flat, it will expand forever and continue to cool from its present temperature (about 2.7^0K), asymptotically approaching absolute zero. The presence of "cosmological constant, Λ" could either accelerate its expansion, or, possibly close the universe. If it is closed, it will expand to a maximum size in another 1 to 500 billion years, then recollapse to an arbitrarily small size and unendingly higher temperatures somewhat like a mirror image of its past expansion.

What about the future of life in the universe? It turns out that the overall picture is bleak, regardless of whether it is open or closed, "freeze" or "fry." A reasonably well-agreed-upon account of both closed and open scenarios has been given by Frank Tipler and John Barrow:[9] In 5 billion years, the sun will become a red giant, engulfing the orbit of the earth and Mars, and eventually becoming a white dwarf. In 40–50 billion years, star formation will have ended in our galaxy and others. In 10^{12} years, all massive stars will have become neutron stars or black holes.[10] In 10^{19} years, dead stars near the galactic edge will drift off into intergalactic space; stars near the center will collapse together forming a massive black hole. In 10^{31} years, protons and neutrons decay into positrons, electrons, neutrinos, and photons. In 10^{34} years, dead planets, black dwarfs, neutron stars, will disappear, their mass completely converted into energy, leaving only black holes, electron-positron plasma and radiation. All carbon-based life-forms will inevitably become extinct. Beyond this, solar mass, galactic mass, and finally supercluster mass black holes will evaporate by Hawking radiation. The upshot is clear: "Proton decay spells ultimate doom for life based on protons and neutrons, like *Homo sapiens* and all forms of life constructed of atoms. . . ."[11]

Now we can return to our key question: *Can Christian eschatology be seen as consistent with either of these scientific scenarios?*

New Testament Debates over the Bodily Resurrection of Jesus

In this paper I will adopt as a working hypothesis or "test case" that interpretation of the resurrection of Jesus which poses the most profound challenges for theology when scientific cosmology is taken seriously: namely the bodily resurrection of Jesus (including the empty tomb traditions) and thus eschatology as the transformation of the universe (the creation) into the new creation. This position is well defended by New Testament scholars and theologians, so that the option to adopt it as a test case for the encounter with cosmology is far from arbitrary or easily avoidable.

The resurrection of Jesus has been given what can be called "objective" and "subjective" interpretations. According to those holding the objective interpretation, something happened to Jesus of Nazareth after his crucifixion, death, and burial such that he is risen, he lives forever with God, and is present to us in our lives. In short, God raised Jesus from the dead. In the objectivist view, the resurrection of Jesus refers to something which happened to Jesus of Nazareth which cannot be reduced entirely to the experiences of the disciples.[12] According to Raymond Brown, "[o]ur generation must be obedient . . . to what *God* has chosen to do in Jesus; and we cannot impose on that picture what we think God should have done."[13]

The subjective interpretation focuses strictly on the experiences as reported in the appearances and empty tomb traditions. These experiences caused the first disciples to know and believe something new about Jesus of Nazareth after his crucifixion. Here, however, language about the resurrection of Jesus is only a way of speaking about the experiences of the disciples and not about purported events in the new life given to Jesus by God after his death and burial.[14] According to Willie Marxsen, "[a]ll the evangelists want to show is that the activity of Jesus goes on. . . . They express this in pictorial terms. But what they mean to say is simply: 'We have come to believe.'"[15]

The objective interpretation of the resurrection of Jesus emphasizes elements of continuity (or identity) and of discontinuity (or transformation) between Jesus of Nazareth and the risen Jesus, holding these in tension by such phrases as *identity-in-transformation*. These elements of continuity include everything about the human person of Jesus: there is at least a minimal element of physical, material, personal and spiritual continuity between Jesus of Nazareth and the risen Jesus. Because of this, most scholars following the objective interpretation use the term *the bodily resurrection* to emphasize the significance of the empty tomb traditions, and thus the inclusion of the physical in the overall meaning of the resurrection.[16]

Scholars who support the bodily resurrection of Jesus connect his resurrection with the general resurrection "at the end of time" and the "new creation" consisting of a "new heaven and earth" tend to overlook the challenge from cosmology. They view the "new creation" as a transformation of the world as a whole and all that is in it; it is a return of the risen Christ to *this* world in order that this world be *transformed* into an eternal world. Curiously, the challenge raised by scientific cosmology to this claim is seldom inspected.

For the purposes of this paper I will work with the "bodily resurrection of Jesus" interpretation of the resurrection texts since it constitutes the "test case/worst case" approach to the conversations with science.

Resurrection, Eschatology, and Cosmology within "Theology and Science": The Surprising Lack of Engagement

Against competing claims that theology and science are necessarily in conflict or in separate worlds, there has been enormous progress over the past fifty years in the approach that places theology and science in dialogue and interaction.[17] Physics, cosmology, evolutionary and molecular biology, and other areas of the natural sciences have been introduced into ongoing theological discussions. Particular attention has been given to the goal of articulating objectively special but noninterventionist divine action in particular by searching for genuine openness in nature (ontological indeterminism).[18] Little attention, however, has been given to the resurrection of Jesus and its eschatological implications in light of the natural sciences.[19] This lack of attention is particularly ironic since the same methodological framework which has played an essential role in making it possible for the field of theology and science to grow so richly prevents us from side-stepping the crucial issues raised by cosmology for Christian eschatology.

To see this we need to summarize two of the central claims of this methodology briefly:[20] 1) a nonreductive, holistic view of epistemology, and 2) an analogy between the methodologies of the sciences and the humanities. Here I will be drawing directly on the pioneering writings of Ian Barbour[21] as well as on those of Arthur Peacocke,[22] Nancey Murphy,[23] Philip Clayton,[24] John Polkinghorne,[25] and many others.

The sciences and the humanities, including theology, can be placed in a series of levels which reflect the increasing complexity of the phenomena they study. In this epistemic hierarchy, lower levels place epistemic *constraints* on upper levels (against two worlds), but upper levels cannot be reduced entirely to lower levels. Thus, physics as the bottom level places constraints on biology. On the other hand, the processes, properties, and laws of biology *cannot be reduced* without remainder to those of physics.

Within this hierarchy, each level involves similar methods of theory construction and testing. Thus theological methodology is analogous to scientific methodology (though with several important differences).[26] This claim is both a *description* of the way many theologians actually work and a *prescription* for progress in theological research. Theological doctrines are seen as theories, working hypotheses held fallibly, constructed through metaphors and models, and tested in light of the data of theology, now including the results of the sciences.

In order to clarify the ways in which science can and should influence theology, I will combine these two ideas—epistemic hierarchy and analogous methodology—into a single framework, as indicated by Figure 6 (for now, paths 1–5). For simplicity I'll limit the conversation to physics and its effects on theology.[27] Thus there are five paths by which the natural sciences can affect constructive theology:

1. Theories in physics can act directly as data which place constraints on theology. So, for example, a theological theory about divine action should not violate special relativity.

2. Theories can act directly as data either to be "explained" by theology or as the basis for a theological constructive argument. Thus t = 0 in standard Big Bang cosmology was often explained theologically via creation *ex nihilo*.

3. Theories in physics, after philosophical analysis, can act indirectly as data for theology. For example, an indeterministic interpretation of quantum mechanics can function within philosophical theology as making intelligible the idea of non-interventionist objective divine action.

4. Theories in physics can also act indirectly as the data for theology when they are incorporated into a fully-articulated philosophy of nature (e.g., that of Alfred North Whitehead). Finally,

5. Theories in physics can function heuristically in the theological context of discovery, by providing conceptual or aesthetic inspiration, etc.

So biological evolution may inspire a sense of God's immanence in nature. For convenience, I will use the symbol, "SRP —> TRP" suggested by George Ellis[28] to indicate these five ways scientific research programs (SRPs) can influence theological research programs (TRPs).

We are now prepared to see clearly why the problem of eschatology and cosmology is forced on us by the same methodological framework which has played an essential role in making it possible for the field of theology and science: Scientific cosmology (i.e., Big Bang cosmology, inflationary Big Bang, quantum cosmology, etc.) is part of physics (i.e., a solution to the field equations of general relativity). Therefore the predictions of freeze or fry must place constraints on and challenge what theology can claim eschatologically. No appeal to contingency, quantum physics, chaos theory, Whiteheadian novelty, emergence, the unpredictibility of the future, or metaphysics alone will be sufficient to solve this problem.[29]

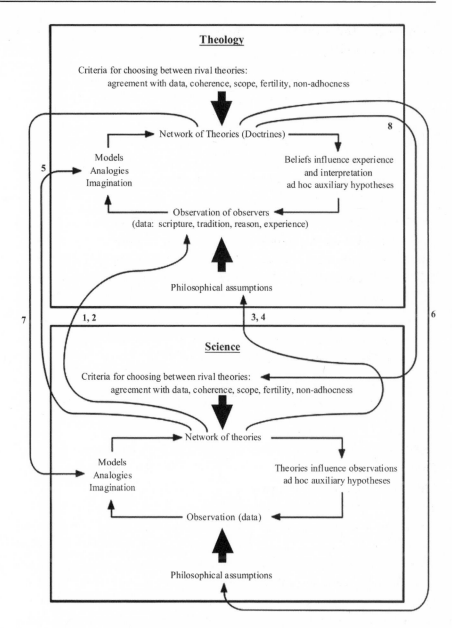

Figure 6: Method of Creative Mutual Interaction

John Polkinghorne is among the very few scholars who have attended in detail to the challenge of cosmology to an eschatology. His work is based on the bodily interpretation of the resurrection of Jesus.[30] Just as Jesus' body was transformed into the risen and glorified body, so the "matter" of this new environment must come from "the transformed matter of this world": "[T]he first creation was *ex nihilo* while the new creation will be *ex vetere* . . . the new creation is the divine redemption of the old. . . . [This idea] does not imply the abolition of the old but rather its transformation."[31] Clues to what the "new heaven and new earth" will be like come from the themes of continuity and discontinuity found in the gospel accounts of the resurrection and in the Eucharist. Moreover, science "may have something to contribute" to our understanding of this transformation: The continuities might lie within the province of science (or, more precisely, what Polkinghorne calls "metascience.")[32]

I will use Polkinghorne's ideas as a starting point. But before we can attempt to proceed, we must develop an extended new methodology in order to address the severe challenge placed by the existing methodology on such attempts as described above.

METHODOLOGY AND GUIDELINES FOR NEW RESEARCH IN SCIENTIFIC COSMOLOGY AND IN ESCHATOLOGY

I see that you believe these things are true because I say them.
Yet, you do not see how.
Thus, though believed, their truth is hidden from you.

—*Dante Alighieri*[33]

If it is impossible, it cannot be true. But if it is true, it cannot be impossible.

It seems clear that if one assumes the current methodology in theology and science and if one assumes the test case, the bodily resurrection of Jesus and an eschatology of cosmological transformation, then one *seems* forced into a direct contradiction with the predictions of contemporary scientific cosmology. To move forward, I propose we expand our methodology to allow for genuine interaction between theology and science. I also believe that an expanded method is, from an informal point of view, already in place. Making it explicit should be helpful for a variety of problems in theology and science, possibly including our current problem.

Let me describe the expanded methodology by referring to three paths in Figure 6.1 (6, 7, 8) which represent the possible influences of theology on the philosophical assumptions which underlie science and/or on individual scientists or teams of scientists in that initial phase of research often called the "context of discovery." I will again use Ellis's suggestion to represent these paths as TRP—>SRPs. At the outset, however, I want to stress that by "influence" I am *not* assuming that theologians speak with some special kind of dogmatic authority. Quite the contrary; the overall context should be an open intellectual exchange between scholars based on mutual respect and the fallibility of hypotheses proposed by either side.

(Path 6): It is now abundantly clear that, historically, theological ideas provided some of the philosophical assumptions which underlie scientific methodology. Historians and philosophers of science have shown in detail how the doctrine of creation *ex nihilo* played an important role in the rise of modern science by combining the Greek assumption of the rationality of the world with the theological assumption that the world is contingent. Together these helped give birth to the empirical method and the use of mathematics to represent natural processes.[34]

(Path 7): Theological theories can act as sources of inspiration in the scientific context of discovery, that is in the construction of new scientific theories. An interesting example is the subtle influence of atheism on Hoyle's search for a "steady state" cosmology.[35]

(Path 8): Theological theories can lead to "selection rules" within the criteria of theory choice in physics. For example, if one considers a theological theory as true, then one can delineate what conditions must obtain within physics for the possibility of its being true. These conditions in turn can serve as motivations for an individual research scientist or group of colleagues to choose to pursue a particular scientific theory.

The asymmetry between theology and science should now be quite apparent: Theological theories do not act as data for science, placing constraints on which theories can be constructed in the way that scientific theories do for theology. This, again, reflects the prior assumption that the sciences are structured in an epistemic hierarchy of constraints and irreducibility. It also safeguards science from any normative claims by theology. Together these eight paths portray science and theology in a much more interactive mode. I suggest calling this "*the method of creative mutual interaction.*" Given this method, we can begin to delineate the conditions needed for real progress in theology and science.

Guidelines for Moving Forward

Given the expanded methodology described above we are prepared to engage in a twofold project: First, following paths 1–5, we construct a more nuanced understanding of eschatology in light of physics and cosmology (indicated by the symbol: SRP —> TRP). Second, following paths 6–8, we search for a fresh interpretation of, or possibly revisions of, current scientific cosmology in light of this eschatology (indicated by the symbol: TRP —> SRP). If such a project is at all successful, it might eventually be possible to bring these two trajectories together at least in a very preliminary way to give a more coherent overall view than is now possible of the history and destiny of the universe in light of the resurrection of Jesus and its eschatological completion in the parousia.

This project is clearly a long-term undertaking, requiring the participation of scholars from a variety of fields in the sciences, philosophy, and theology. How do we then proceed? My sense is that we first need some guidelines that will help point us in a fruitful direction. Using them, we can begin to explore specific ways to enter into the research.

SRP —> TRP

We begin with guidelines for constructive theology in light of contemporary science. Our first four guidelines deal with overall philosophical and method-ological issues (paths 1–5).

Guideline 1:   Rejection of two philosophical assumptions about science: the argument from analogy and its representation as nomological universality

The first guideline deals with the fundamental challenge physical cosmology poses to the kind of eschatology we are considering here: namely, one based on our choice of the "hardest case"—the bodily resurrection of Jesus. In bare form, the challenge is start: If the predictions of contemporary scientific cosmology come to pass, then the parousia will not just be "delayed," it will never happen. And if this is so, then the logic of Paul in 1 Corinthians 15 is then inexorable: If there will never be a general resurrection, then Christ has not been raised from the dead, and our hope is in vain. The challenge can also be seen as coming from theology to science: If it is in fact true that Jesus rose bodily from the dead, then the general resurrection cannot be impossible. This must in turn mean that the future of the universe will not be what scientific cosmology predicts.

We seem to be at loggerheads. How are we to resolve this fundamental challenge? My response is that the challenge is not technically from science but from a philosophical assumption which we routinely bring to science, namely that scientific predictions hold without qualification. It is quite possible, however, to accept a very different assumption about the future predictions of science while accepting all that science describes and explains about the past history of the universe. The first step is deciding whether the laws of nature are descriptive or prescriptive and, as Bill Stoeger argues, science alone cannot settle the matter.[36] A strong case can then be made on philosophical grounds that the laws of nature are descriptive. The second step is to claim on theological grounds that the processes of nature which science describes in terms of the laws of nature are actually the result of God's ongoing action as Creator and not of nature acting entirely on its own. The regularity of natural processes is ultimately the result of God's faithfulness, even if God bequeaths a significant degree of causal autonomy to nature. (Here one thinks of the traditional debates between Augustinian and Thomistic interpretations of natural causality: the former viewing nature as highly contingent upon God and natural events as "omnimiraculous," and the latter viewing nature as gifted by God with relative autonomy and a high degree of causal efficacy.) Finally, if this is so, and if God is free to act in radically new ways not only in human history but in the ongoing history of the universe (which of course God is!), then the future will not be what science predicts. Instead it will be based on a radically new kind of divine action which began with the resurrection of Jesus, and this new act of God cannot be reduced to, or explained by, the current laws of nature, that is, by God's action in the past history of the universe.[37]

In short, we could say that the freeze or fry predictions for the cosmological future might have applied had God not acted in Easter and if God were not to continue to act to bring forth the ongoing eschatological transformation of the universe. Because of Easter and God's promise for its eschatological completion, however, the freeze or fry predictions will not come to pass.

Guideline 2: Eschatology should embrace methodological naturalism regarding the cosmic past and present.

Any eschatology which we might construct must be "scientific" in its description of the *past* history of the universe. More precisely, it must be constrained by methodological naturalism in its description of the past: it should not invoke God in its explanation of the (secondary) causes, processes and properties of nature.[38] This guideline separates this proposal as sharply as possible from such approaches as "intelligent design" in so far as they are critical of the physical and/or biological sciences for not including divine agency in their mode of explanation.

Guideline 3:   "Relativistically correct eschatology": constructing eschatology in light of contemporary physics (paths 1, 2).

Although we will set aside the predictions Big Bang offers for the cosmic future, we must be prepared to reconstruct current work in eschatology in light of Big Bang cosmology and contemporary physics—primarily specifically relativity and quantum physics—and thus what cosmology tells us about the history of the universe, following paths 1 and 2. I will refer to this project as the attempt to construct a "relativistically correct Christian eschatology."

Guideline 4:   Big Bang and inflationary cosmology as a "limit condition" on any revised eschatology (path 1): the material argument.

This guideline follows path 1 by stating that standard and inflationary Big Bang cosmologies, or other scientific cosmologies (such as quantum cosmology), place a "limiting condition" on any possible eschatology. All we know of the history and development of the universe and life in it will be data for theology.

Guideline 5:   Metaphysical options: limited but not forced.

In revising contemporary eschatology there are various metaphysical options from which we may choose; they are not determined by science. These options include physicalism, emergent monism, dual-aspect monism, ontological emergence, and panexperientialism (Whiteheadian metaphysics).[39]

We must now begin the enormous job of revisiting our understanding of eschatology in a way that takes up and incorporates all the findings of science, and particularly scientific cosmology, without falling back into the philosophical problem guideline 1 is meant to address.

Guideline 6:   "Transformability" and the formal conditions for its possibility (the "such that"[40] or "transcendental" argument).

Our starting point is that the new creation is not a replacement of the old creation, or a second and separate creation *ex nihilo*. Instead, God will transform his creation, the universe, into the new creation *ex vetere*, to use Polkinghorne's phrase. It follows that God must have created the universe *such that it is transformable*, that is, that it can be transformed by God's action. God must have created it with precisely those conditions and characteristics which it needs as preconditions in order to be transformable by God's new act. Moreover, if it is to be transformed and not replaced, *God must have created it with precisely those conditions and characteristics which will be part of the new creation*. Since science offers a profound understanding of the past and present history of the universe (guidelines 2, 3), then science can be of immense help to the theological task of understanding something about that transformation if we can find a way to identify, with at least some probability, these needed conditions, characteristics and preconditions. I will refer in general to these conditions and characteristics as "elements of continuity."

Guideline 6 can be thought of as a *transcendental* or "such that" argument.[41] A simple analogy would be that an open ontology can be thought of as providing a precondition for the enactment of voluntarist free will, but certainly not the sufficient grounds for it. Science might also shed light on which conditions and characteristics of the present creation we do *not* expect to be continued into the new creation; these can be called "elements of discontinuity" between creation and new creation. Thus physics and cosmology might play a profound role in our attempt to sort out what is truly essential to creation and what is to be "left behind" in the healing transformation to come.

This guideline gives to the terms *continuity* and *discontinuity*, found in the theological literature on the resurrection of Jesus, a more precise meaning and a potential connection with science. With it in place we can move eventually to a material argument and ask just what those elements of continuity and discontinuity might be.

Guideline 7:   Continuity within discontinuity: inverting the relationship.

Closely related to the previous guideline is a second formal argument about the relative importance of the elements of continuity and discontinuity. So far in theology and science, discontinuity has played a secondary role within the underlying theme of continuity in nature as suggested by the term *emergence*. Accordingly, irreducibly new processes and properties (discontinuity) arise within the overall, pervasive, and sustained background of nature (continuity). Thus biological phenomena evolve out of the nexus of the physical world, the organism is built from its underlying structure of cells and organs, the mind arises in the context of neurophysiology, and so on. Now, however, when we come the resurrection and eschatology, I propose we *invert* the relation: the elements of "continuity" will be present, but within a more radical and underlying "discontinuity" as is denoted by the *transformation* of the universe by a new act of God *ex vetera*. With this inversion, discontinuity as fundamental, signals the break with naturalistic and reductionistic views such as "physical eschatology," while continuity, even if secondary, eliminates a "two-worlds" eschatology.

This has important implications on our search for candidate theories. It *eliminates* "noninterventionist objective special divine action" as a candidate since it does not involve a transformation of the whole of nature. Indeed, these approaches presuppose that it is the continual operation of the "usual" laws of nature which make objective special divine action possible without the need for their violation or suspension. But the bodily resurrection of Jesus directs us towards a much more fundamental view: the radical transformation of the background conditions of space, time, matter and causality, and with this, a permanent change in at least most of the present laws of nature.[42]

TRP—>SRP

Our project also involves the question of whether such revisions in theology might be of any interest to contemporary science—at least for individual theorists who share eschatological concerns such as developed here and are interested in whether they might stimulate a creative insight into research science (paths 6–8).

Guideline 8:   Theological reconceptualization of nature leading to philosophical and scientific revisions (path 6).

Here we move along path 6 in discovering whether a richer theological conception of nature both as creation *and* as new creation can generate important revisions in the *philosophy of nature* that currently underlies the natural sciences, the philosophy of space, time, matter and causality in contemporary physics and cosmology.

Guideline 9:   Theology as suggesting criteria of theory choice between existing theories (path 7).

We can also move along path 7 to explore philosophical differences in current options in theoretical physics and cosmology. The theological views of research scientists might play a role in selecting which theoretical programs to pursue among those already "on the table" (for example, the variety of approaches to quantum gravity).

Guideline 10:   Theology as suggesting new scientific research programs (path 8).

Finally we can move along path 10 and suggest the construction of new scientific research programs whose motivation stems, at least in part, from theological interests.

In closing this section I want to stress once again that all such programs in science would have to be tested by the scientific communities (what is often called "the context of justification") without regard for the way theology or philosophy might have played a role in their initiation ("the context of discovery").

SUGGESTIONS FOR RESEARCH PROGRAMS IN THEOLOGY AND IN SCIENCE

SRP —> TRP
Reconstructing Christian Eschatology as "Transformation"

In order to move ahead and in light of our guidelines, I suggest the following directions for research: Continuities, Discontinuities, and Their Preconditions for the Transformation of the Creation. Following guidelines 6 and 7 in particular, we start by focusing on continuities, discontinuities and their preconditions that are part of that transformation. These may be found in certain suggestive eschatological hints gleaned from the resurrection of Jesus and the reign of God depicted in the New Testament and its glimpses in the living body of Christ, the Church, and keeping clearly in mind the apophatic character of this material.

Continuities and discontinuities (G7)

Hints of continuity from the a) *resurrection of Jesus*: he could be touched, he could eat, break bread, be seen,[43] heard and recognized. These instances of "realized eschatology" are suggestive of an extended "domain" of the new creation propleptically within the old, a domain which ceases with the "ascension," but which, when present, includes Jesus, the disciples, and their surroundings.[44] Hints of discontinuity: these encounters with the risen Lord break with a mere resuscitation and with normal limitations on "physicality."

Hints of continuity from the b) *reign of God* in the New Testament and in the church: the "new creation" will include persons-in-community and their ethical relations.[45] Hints of discontinuity: in the "reign of God," it will "not be possible to sin," compared with the present creation in which it is "not possible not to sin," to use Augustine's apt formulation.

Hints of continuity from the problem of *personal identity* between death and the general resurrection in historical and contemporary theology: i) Paul's analogy of the seed (1 Cor 15:35ff.); ii) numerical, material and/or formal continuity between death and general resurrection in historical Christian thought.[46] Hints of discontinuities: though death awaits all people now, the general resurrection, like that of Jesus at Easter, will not be a resuscitation; Paul's fourfold contrast (1 Cor 15:42ff.).

Preconditions for these continuities and discontinuities (G6)

Next, following guideline 6, we look for epistemically "prior" elements of continuity by moving "down" the disciplinary hierarchy, focusing on those characteristics which make possible the elements of continuity touched on above. For now, I will set aside the many levels to be found in the social, psychological, and neurosciences, and move directly to physics. A central theme underlying a) and b) above is temporality; thus we would expect that time as understood by physics is not only a characteristic of this universe as God's creation, but that it will in some ways be a characteristic of the new creation. Yet we would expect that in the new creation, our experience of temporality will no longer be marred by the loss of the past and the unavailability of the future. So there will be an element of discontinuity during the transformation as well. We could make a similar case for ontological openness as the enduring precondition for persons-in-community to act freely in love, so that this too might be an element of continuity during the transformation of the universe. Other examples include ontological relationality/holism, the role of symmetries/conservation laws, and so on.[47] Presumably mathematics, too, will be an element of continuity, and in this case, perhaps without discontinuity.[48]

The Theological Research Program (TRP): Reconstructing eschatology

The next step is to undertake a reconstruction of Christian eschatology in light of these arguments and following guidelines 3 and 4, in light of contemporary science and cosmology, at least regarding the past history and present state of the universe. This will clearly require extensive research far beyond the limits of this paper. Nevertheless, some hints of the way forward can be found in focusing briefly on one aspect of that project: the relation of time and eternity.

A crucial argument shared widely among contemporary theologians[49] is that eternity is a richer concept of temporality than timelessness or unending time. In essence, eternity is the source of time as we know it, and of time as we will know it in the new creation. Eternity is the fully temporal source and goal of time. Barth calls it "supratemporal." Moltmann calls it "the future of the future," and Peters refers to the future as coming to us (*adventus*) and not merely that which tomorrow brings (*futurum*). Pannenberg claims that God acts propleptically from eternity: God reaches back into time to redeem the world, particularly in the life, ministry, death and resurrection of Jesus. In this approach the relation between "time and eternity" is modeled on the relation of the finite to the infinite. Here the infinite is not the negation of the finite (as in the Platonic/Augustinian view of eternity as timelessness); instead the infinite includes and yet ceaselessly transcends the finite.

In my opinion, this view of eternity includes at least five distinct themes:

1. copresence of all events: I define the term *copresence* to mean that distinct events in time are nevertheless present to one another without destroying or subsuming their distinctiveness;

2. "flowing time": each event has what I will call a "past/present/future structure" (or "ppf structure"); this structure is that of an "inhomogeneous temporal ontology";

3. duration: each event has temporal thickness in nature as well as in experience; events are not point-like present moments lacking an intrinsic temporal structure;

4. global future: there is a single global future for all of creation so that all creatures can be in community;

5. prolepsis: the future is already present and active in the present while remaining future, as exemplified by God's act in raising Jesus from the dead.

Please note that the combination of flowing time and copresence in eternity means that all events can be "simultaneously present" and available to each other (á Boethius), and yet each event retains its own unique identity. Thus *in the new creation, "flow" keeps "copresence" from reducing to "nunc" while "copresence" keeps "flow" from reducing to a stream of isolated moments.*

Our guidelines now lead to three questions: 1) Which of these themes are already present in creation and thus are elements of continuity in its transformation into the new creation? 2) Which themes are not yet present in creation but instead represent elements of discontinuity, emerging only in the new creation? 3) Regarding the latter, does the universe at present include the preconditions for the possibility of their coming to be in the new creation? The answer to these questions will require a careful discussion of time in physics and cosmology. It will also require us to reformulate these theological themes in terms of our current understanding of time as drawn from twentieth-century physics and cosmology.

Though the reformulation has only begun, I can nevertheless anticipate possible responses to the three questions. To the first, I would argue that *flowing time and duration* are objective features in nature, though this is debatable. Time in special relativity is subject to conflicting interpretations (e.g., "block universe" as well as "flowing time"). Moreover, lacking copresence, "flowing time" for us means an isolated present with a vanishing past and a not-yet realized future. Following path 3, I would introduce a "copresent flowing time" interpretation of special relativity into the theological discussion of time and eternity. Duration is a harder problem, since contemporary physics assumes a point-like or durationless view of time. I believe, however, that a case can be made for duration in nature by drawing on Pannenberg's arguments.[50]

To the second and third questions, I would identify what I am calling "copresence," prolepsis, and the global future as elements of discontinuity, and thus search physics for their preconditions in nature. For example, the transformation of flowing time into copresent flowing time would seem plausible if one could argue that the inhomogeneous ontology of flowing time does not logically exclude the possibility of distinct temporal events being copresent. Preconditions for prolepsis could include backward causality, violations of local causality, and violations of global causality. Finally, a global future, while excluded by special relativity, is theoretically possible in general relativity where the topology of the universe is contingent on the distribution of matter.

The theological task will now be to reconstruct eschatology with these insights in mind and with our theological concepts reformulated in light of contemporary physics. For example, time and space are treated as independent quantities in classical physics. Similarly, theological treatments of eternity and omnipresence typically take for granted independent treatments of time and space. But time and space are placed in a complex interrelationship in special relativity and further linked with matter in general relativity. Our task then will be to reformulate such theological categories as eternity and omnipresence in light of special and general relativity with an eye to "copresent flowing time" and time as duration. Similar theological reconstructions will hold for the treatment of time and space in quantum mechanics, and so on.

TRPs –> SRPs
Directions for Potential Research Programs in Physics and Cosmology

Our methodology of mutual interaction includes a second agenda: Explore ways in which a revised eschatology as suggested above leads to a revised philosophy of nature as it underlies science, to criteria of theory choice among current theories in science, and to the construction of new scientific research programs.

The first project, revising eschatology in light of science, is still in its very early stages. Until more has been accomplished, we might pursue the second agenda by a more limited approach: begin with our existing eschatology and the *same* elements of continuities, discontinuities and their preconditions as listed above; and ask now what SRPs they might suggest. Once again, since temporality is such a predominant theme here, we could start with the theme of "time and eternity,"[51] but now we will explore its implications for current physics and cosmology in two ways: 1) We begin with those aspects of temporality in nature, such as "flowing time" and "duration," which constitute elements of continuity and of discontinuity in respect to the eschatological transformation of the world. This time, however, the analysis should lead to interesting questions about time in current physics. 2) We will also consider aspects of temporality that physics may have overlooked but which, from the perspective of eschatology, might be expected to exist at present, such as copresence, prolepsis, and global future. If physics were reconsidered with them in mind, could it generate concrete suggestions for research programs in physics?

Flowing time and duration vs. special relativity and quantum mechanics

Let us return to the debate discussed above regarding "flowing time" versus the "block universe" in special relativity. Here the absence of any physical meaning for absolute simultaneity (i.e., a universal and unique present moment) seems to challenge most approaches to the idea of "flowing time." Can we construct a new interpretation of special relativity which is consistent with "flowing time" but which avoids these problems, as William Lane Craig and others suggest?[52] I will label this *SRP(1)*. Alternatively, scholars such as John Lucas[53] suggest we revise special relativity to support a flowing time interpretation over a block universe interpretation, which I will designate as *SRP(2)*.

The standard formalism of quantum mechanics presupposes flowing time, but the interpretation of this formalism is highly debated. The Copenhagen interpretation, although widely accepted among physicists and philosophers of physics, does presuppose flowing time.[54] However, given such internal problems as the "measurement problem" and given the vitality of competing interpretations one cannot rely on the Copenhagen interpretation to warrant an "objective arrow of time" in nature. One alternative would be to consider constructing a new and more competitive interpretation of quantum mechanics which more definitively supports flowing time *(SRP[3])*. Another would be to explore modifications of the actual formalism of quantum mechanics which support flowing time, such as nonlinear or stochastic versions of the Schrödinger equation *(SRP[4])*.

Finally, one might pursue Pannenberg's arguments that duration should be found in nature by searching for mathematical ways to represent time as duration, such as set theory or quantized time, and then by exploring their implications for research physics *(SRP[5])*.

Copresence, prolepsis, and global future in physics and cosmology

We now turn to the preconditions for the possibility of continuity: copresence, global future, and prolepsis. I will touch on the first two briefly, leaving more room for extended comments on the third precondition.

By *copresence* I mean the "temporal" (or, perhaps more accurately, "transtemporal") presence to one another of distinct events in time, a presence which does not destroy or subsume their distinctiveness (e.g., what I have termed their unique ppf structure).[55] It is challenging to think of positive preconditions on the nature of flowing time which make its transformation into copresence possible. We might, however, be able to formulate a "negative" precondition: There should be nothing about flowing time that makes its transformation into copresent flowing time intrinsically impossible or the concept of such a transformation entirely unintelligible.

By *global future* I mean that all events lie in the causal future of some future set of events. A common theme in Christian eschatology is the unity of the future: The creation will be transformed into a single global domain so that all creatures in the new creation can be in community. Is there a unique global causal future which could serve as a precondition for the transformation by God of the predicted future into the eschatological future of the new creation?

According to SR, there is no unique global causal future. The causal futures of any two events P and Q will share some events in common, but there will always be other events which lie either in the causal future of P but not of Q, or of Q but not of P. GR, however, provides a possible precondition for the theme of the eschatological global future: It shows that the topology of the universe is contingent on the distribution of matter, and it allows for a variety of topologies for the future, including ones in which geodesics do not separate to arbitrary distances.

The theological meaning of *prolepsis* is that God acts from the eternal future in the present. The primary instance is God's act in raising Jesus from the dead and, in turn, the ongoing transformation of creation into new creation. Clearly the theological meaning of "future" is distinct from the scientific meaning; yet the attempt here is to find ways to bring them into a more fruitful relationship. Thus there may be features in physics and cosmology which were created by God *ex nihilo* such that they might function as physical preconditions for the possibility of proleptic action of God; they might do so by pointing to or indicating the intelligibility of "reverse causality" already present in nature. In essence, the causal structure of the universe might be more subtle than the "arrow of time" discussion allows; it might not be entirely inconsistent with the idea that the future transformation of the universe by God effects the present through reverse causal processes already at work in nature. For brevity, I will simply list examples of how these features have been discussed in physics.

Backward causality in time-symmetric formulations

The equations of electromagnetism are "time symmetric:" Mathematical solutions to these equations can describe the propagation of light forward in time, as we expect from experience, or backward in time, which is contrary to experience. Nevertheless mathematical models of the propagation of photons based on various combinations of such forward and backward solutions (or, as they are technically called, "retarded" and "advanced" solutions, respectively) have been explored for decades following the catalytic papers by Richard Feynman and J. A. Wheeler.[56] They illuminate several technical problems in electromagnetic theory, and they also provide a very simple model of how "the future could affect the past" without propagation at speeds greater than light.

Ironically, Fred Hoyle's "steady state cosmology" was the result of a modification of GR based, in part, on the idea of advanced and retarded gravitational potentials.[57] Although the Big Bang models based on GR "won the day" in the 1960s, the time symmetric approach begun by Hoyle still has scientific advocates today.[58] Thus a direction for research programs motivated in part by theological views about prolepsis and designed along these lines is at least in principle viable.

Violations of local causality (propagation at speeds greater than light)

The expansion of the universe now appears to be accelerating instead of decelerating the way the standard open Big Bang models predicted. The acceleration might be explained by adding the so-called "cosmological constant, Λ," to Einstein's field equations. This addition represents a "pressure" term typical of an incompressible fluid and suggests a "physical" reason for the increasing acceleration of the universe's expansion.[59]

What is surprising is that Einstein's equations with the inclusion of Λ allow, at least formally, for the propagation of sound exceeding the speed of light and thus possibly going backwards in time. Such backward waves represent one possible way that causality within small regions of space-time might be violated. In a recent work, George Murphy has discussed the potential significance of such backward waves for eschatology.[60] (Backward waves are also allowed in the de Sitter static space-time.[61])

Violations of global causality (closed paths in space-time)

It is well known that Gödel's universe allows for timelike paths (paths along which matter can move) that are closed but that do not violate the speed of light.[62] Essential singularities in space-time allow passage from one portion of our universe (e.g., "the future") to another portion (e.g., "the past")[63] or another universe. To pursue these questions requires the use of global techniques in the analysis of such topics as time/space orientability, the chronology condition, the causality condition, the future/past distinguishing condition, the strong causality condition, the stable causality condition, the existence/nonexistence of a Cauchy surface (or a partial Cauchy surface); etc.[64]

SRP (path 6): Regarding copresence, global future, and prolepsis we may offer the following SRP guidelines 8–10: Consider and develop ways in which these and other features in physics and cosmology point to the complex character of time and time's arrow as indicative of the preconditions for copresence, prolepsis, and the global future.

Conclusion

Many distinct areas in biblical studies and systematic theology require us to give central attention to eschatology, including the bodily resurrection of Jesus and its implications for the new creation, the problem of suffering both in humanity and in the whole sweep of biological evolution, the "crucified God" as a response to suffering and its demand for an eschatology of new creation, indeed the very meaning of a relational understanding of the Trinitarian doctrine of God, not to mention the program of theology and science itself. But eschatology of this sort, with its cosmic horizons unavoidable, faces the severe challenge of the scientific prognosis for the far future: freeze or fry. Even if scientific cosmology changes, as indeed it will do (and is doing), the strictly scientific changes will never be capable of providing on their own a basis for such an eschatology.

Instead of challenging science, we recognize that we need not make the strictly philosophical assumption that what science predicts must come to pass. We can think about the future of the universe in theological terms, through terms which have been heavily engaged and reconstructed in light of what the sciences tell us validly about the past and present of the universe. Such a reconstructed eschatology might, in turn, offer new insights about the present creation which could be fruitful for those engaged in scientific research. In sum, we are asking what an expanded scientific conception of nature would be like if we inherited from eschatology instead of creation theology, and what ramifications this might have for current science.

Taken together—the reconstruction of eschatology in light of science but based on its theological and biblical resources, and the indication of directions for potential research in science from this new eschatological perspective on the universe at present—represents an instance of the methodology of creative mutual interaction between theology and science. The extensive development of this and other representations of this interaction is currently in process. The value of the results of this interaction can only be judged when those results have been more fully articulated, but I do believe that in principle the setting out of such a new methodology is in itself of lasting value and provides an open invitation to a variety of scholars and a diversity of theological and scientific interests and views.

Chapter 7

Time and the Physics of Sin

Hugh Ross

Any discussion about God, time, and eternity is limited by inconsistent and incomplete definitions of time and temporality. These limitations arise from the fact that all humanity's temporal experiences are confined to a single time dimension in which time can neither be stopped nor reversed. No mere human can get outside of our time dimension to observe objectively all its properties and, thus, arrive at a complete definition.

People can only experience temporal phenomena (cause and effect, emotions, relationships, reasoning, prayer, and all other mental and spiritual activities) along the single time dimension that makes up the space-time manifold (or surface) of the universe. Thus, a person easily falls into the trap of equating "temporality" with "time." They are not exactly the same, and the Bible corrects this mistake.

The Bible (uniquely among all other "holy" books) teaches that when God created the universe he created not just matter and energy but space and time as well. This fact is affirmed in the context of modern cosmology, which says the beginning of the universe is the beginning of length, width, height, and time (plus six other space dimensions that stopped expanding when the universe was only a fraction of a second old). According to Romans 8 and Revelation 20–22, the universe will end—and be replaced—when God's purposes for it have been fulfilled. In other words, the Bible claims that the time dimension in which we experience all temporal phenomena had a definite beginning and will have a definite ending.

From this biblical perspective, temporal phenomena are not limited to the time dimension along which the physical universe unfolds. Scripture reveals that the persons of the Godhead related to one another before the beginning of time and exercised causality before time's beginning. For example, God caused the creation *ex nihilo* of the entire physical universe (Gen 1:1, John 1:1-3, Col 1:15-17, Heb 11:3); God conferred grace (2 Tim 1:9); and God prepared hope (Titus 1:2) even before creating time. The Bible also says that after the universe—after time as we know it—ceases to exist, redeemed humans will still relate to God, to one another and to angels in a far more expansive, fulfilling, and rewarding way than is possible in cosmic time. Our capacity for creative expression and for all manner of emotional, intellectual, and spiritual activities will be greatly expanded in his "new creation." No longer will our temporal capacities and experiences be confined to a single time dimension.[1]

Mathematics can be helpful in demonstrating that temporal phenomena cannot, in principle, be limited to the universe's time dimension. Just as spatial boundaries are superceded as one moves from a single space dimension to two, to three, etc., so also temporal limitations are hurdled as one moves from a single time dimension to two, to three, etc. Illustrations of extradimensional capacities and phenomena appear in chapter seven of *Beyond the Cosmos*, which was included in the syllabus presented at the conference.[2]

The crucial point is that God's existence transcends the dimensions of space and time. According to the Bible, he can create and remove space-time dimensions at will. Therefore, God's "temporal" capacities and activities may take place trans-dimensionally, or in some other way only roughly equivalent to extradimensionality. But at least we can contemplate the reality which extradimensions (or their equivalent) open up.

As an astronomer and a Christian, I am delighted to report that the latest cosmological research supports the biblical notion that time had a beginning *and* that temporal phenomena, as science defines them, preceded the beginning of cosmic time. As the following pages present, astrophysics now attests that the biblical worldview of God, time, and eternity matches not just a preponderance of evidence, but a body of evidence that takes observers beyond all reasonable doubt.

COSMOLOGY AND SCRIPTURE AGREE ON TIME'S ORIGIN

More than three thousand years before any scientist or scholar developed a viable, *testable*, cosmological model, authors of the Bible wrote about the fundamentals of what today is called the Big Bang theory. Several dozen variants of the Big Bang are currently in contention, but all of them include these three foundational concepts: 1) a "singularity" beginning—a unified beginning of matter, energy, space, and time; 2) continual expansion of the cosmos from the creation event; and 3) progressive cooling as time and expansion continue.

Many Bible verses describe the singular beginning,[3] but the most familiar is Genesis 1:1 (AV), which says, "In the beginning God created the heavens and the earth." The Hebrew word for "created" (*bāra'*) means, in this context, "to bring into existence something brand new, something that didn't exist before."[4] And the expression for "the heavens and the earth" (*shāmayim 'eres*) refers to all matter and energy and even the space-time dimensions along which matter and energy are distributed.[5] Hebrews 11:3 records that the universe we can detect was made from that which we cannot detect. John 1, Colossians 1, and numerous other passages offer more detailed accounts of the Big Bang "singularity."

Meanwhile, the space-time theorems of general relativity establish that if the universe has mass and if general relativity indeed holds true, a cosmic beginning—not just of matter and energy, but also of space and time—is inescapable. (Note that the space-time theorems have now been generalized to include all the inflationary Big Bang models, not just the original standard Big Bang model.[6])

According to Oxford's Roger Penrose, general relativity currently ranks as the most exhaustively tested and proven principle in all of physics.[7] So, if the universe contains mass (a bathroom scale convinces most skeptics that it does) and if general relativity is reliable (which it is), one can say with some certainty that the Big Bang beginning, or cosmic singularity, correctly depicts physical reality.[8]

This evidence for cosmic creation testifies specifically to the existence of a *transcendent* Creator, one who exists beyond the boundaries of matter, energy, space, and time. It also speaks of a Creator who fine-tunes the universe in order for life, and specifically human life, to exist. The gods of other religions seemingly create from *within* space and time. The God of the Bible creates from *outside* cosmic space and time. Both science and the Scriptures declare that space and time, not just matter and energy, had their beginning in the finite past.

The Bible says even more about the continual expansion of the universe than it does about the beginning. Five different Old Testament authors— Job, King David, Isaiah, Jeremiah and Zechariah—write about this characteristic expansion.[9]

The psalmist makes the point that the universe has expanded like an unfurling tent. I like to point out that the physical reality of the tent is the surface of the tent. Likewise, cosmologists recognize that matter and energy are distributed along the four expanding dimensions that comprise the surface of the universe.[10] (Six of the initial ten dimensions unfurled to only 10^{-35} meters).

Job in particular makes the point that God alone is responsible for stretching out the heavens. Job's words receive potent confirmation in recent research. First, a paper by Lawrence Krauss[11] predicts the establishment of *two* factors governing the expansion of the universe: mass density which (due to gravity) would tend to slow down the expansion; and the space energy density, which would tend to speed it up.

Krauss claims that this discovery represents "the most extreme fine-tuning problem known in physics." By "problem" he means that in order to explain the possibility of the existence of physical life at any time in the history of the universe, the value of the mass density of the universe (the gravity factor governing the expansion) must be fine-tuned to better than 1 part in 10^{60}. And the space energy density would need to be fine-tuned to better than 1 part in 10^{120}.[12] This number represents the most extreme fine-tuning known in physics.

Of the many (40+ to date) different characteristics of the universe known to require fine-tuning for the possibility of life's eventual existence, these two—the (gravity-determining) mass density and the space energy density—top the list as those with the greatest measurable fine-tuning.[13] So when Job says that God alone stretches out the heavens, he points to a Creator of unimaginably great power and attention to detail. Apparently, human beings are intended to exist.

As for the universe's getting colder and colder, several verses in the Bible address the phenomenon, at least in an indirect way.[14] Probably the most explicit one is Jeremiah 33:25, which declares the fixity of the physical laws that rule the creation. Romans 8 adds that the law of decay is something the entire creation must endure until "the adoption of sons." Until the redemption of God's people and the new creation, the entire universe is subjected to this law of decay, an apparent (to a physicist especially) reference to the second law of thermodynamics, which is intimately tied to the four fundamental forces of physics.

Through these and many other passages, Scripture suggests that the laws of physics are fixed laws. The deduction is as follows: If the physical laws are fixed, thermodynamics are fixed. According to thermodynamics, an expanding chamber always cools with expansion. Compress the chamber, and the temperature of the air in that chamber heats up. Automobile engines operate by this principle. And it applies to everything in the universe. Given the constancy of physics and universality of thermodynamics, the Bible affirms (indirectly but surely) the continual cooling of the universe as it expands.

COSMOLOGY AND SCRIPTURE AGREE ON TIME'S END

Robert Russell and John Polkinghorne have already addressed in this conference a significant and perplexing problem with this Big Bang scenario: it exposes humanity to what seems an unfortunate doom. Having established that the Big Bang theory is a thoroughly biblical concept, we must acknowledge with Russell and Polkinghorne that we are stuck with either a freezing or a fiery end.

This dreadful problem is addressed in some detail in a paper by Lawrence Krauss and Glen Starkman published in *The Astrophysical Journal*—perhaps the most philosophical piece ever to appear in that esteemed periodical.[15]

Krauss and Starkman begin by reviewing the solidity of recent cosmological findings—first, that the universe contains sufficient mass to bring about (under gravity's influence) a slowing of its rate of expansion; and second, that a factor called the "space energy density" currently dominates the mass density in governing the expansion of the universe.

Most people are familiar with the workings of mass and gravity: Under gravity's effect, two massive bodies tend to attract one another, and the closer they are to each other the more powerfully they attract. As the universe emerged from an infinitesimal volume and expanded very rapidly from the creation event, its mass worked powerfully to slow down the expansion. But as the universe has continued to expand and age, the effectiveness of the mass in slowing down the expansion has progressively diminished. So as the universe grows older, and hence bigger, gravity has a weaker and weaker (slowing) effect on the expansion.

The space energy density effect is much harder to grasp. If it had not been verified by a variety of independent experiments, it might seem too strange to be true. Perhaps the best way to illustrate this effect would be to compare (or contrast) it to an elastic band. The space energy density is roughly opposite to the workings of an elastic band. The more an elastic band stretches, the more energy it gains to propel its *contraction*. By contrast, the more the space-time fabric of the universe stretches, the more energy it gains to propel its *expansion*.

Thus, the behavior produced by the space energy density is opposite to that which comes from the mass density (and gravity): When the universe is young and relatively small, the gravity factor is strong in slowing expansion and the space energy density factor is weak in its capacity to propel expansion. But as the universe gets older and bigger, more and more energy becomes available (as described by the space energy density term) to generate expansion.

Astronomers have now confirmed in some detail that the cosmic transition from a slowing expansion to an accelerating expansion occurred roughly 7 billion years ago.[16] In other words, for the past nearly 7 billion years, the universe has been picking up speed rather than slowing in its expansion.

The paper by Krauss and Starkman looks at the consequences for life in an ever-expanding, and ever more rapidly expanding universe. Accelerated expansion is particularly bad news for observational astronomers. Already the universe is expanding so rapidly that the most distant objects in the universe are moving outward at nearly the velocity of light, which means they are on the verge of becoming invisible to astronomers using earth-based telescopes.

As the universe continues to age and to expand with increasing rapidity, more and more objects will be moving away from earth at velocities greater than light's and thus will become invisible to astronomers.[17] How discouraging for an astronomer today but even more so for astronomers of the future! Even with the aid of space telescopes powerful enough to "see" to the theoretical limits of the universe, astronomers will discover (thanks to the space energy density term) that those theoretical limits are beginning to close in on their view. If the sun and earth endured, the sun would someday be moving away from earth more rapidly than the velocity of light, and then earth would no longer receive the sun's heat or light. However, the sun will exhaust all its nuclear fuel long before this happens.

The situation grows worse yet. As the expansion continues and accelerates, star formation will cease. The bits and pieces of matter that coalesce (by gravity) to make stars, will someday fly apart from each other at such a rapid rate that this condensation can no longer occur. Furthermore, because existing stars have a finite life span (at most about 100 billion years—our star, about 9 billion years) the universe will eventually become devoid of luminous stars. And without luminous stars, life is impossible.

In fact, heat flow will eventually diminish to such a degree that metabolism will cease. Proteins will be unable to fold, and the heat flow from hot bodies to cold bodies will become so feeble that metabolic reactions will no longer take place. With that, all physical life must die, and, therefore, all physical consciousness will end.

As Krauss and Starkman declare, an accelerating cosmic expansion inevitably dooms all life in the universe, whether life on planet Earth or life elsewhere.[18] This article most likely ranks as the most depressing paper published to date in *The Astrophysical Journal*. It acknowledges no hope, no destiny, only despair and doom for the cosmos.

The Bible presents—in one respect—the same view of humanity's future. The world and the universe will not endure. Both Old and New Testament passages refer to a time when all the heavens and the earth (echoing Gen 1:1) will be "rolled up like a scroll" (Isa 34:4) and everything will be "removed from its place" (Rev 6:14). In the words of Jesus, as recorded in all three Synoptic Gospels, "heaven and earth will pass away."[19] John's unique vantage point (in Revelation) provides a preview of the moment when Jesus' words are fulfilled.

The biblical story does not end there, however. It ends with a resounding affirmation of humanity's hope. Graciously, in his awareness of our tendency to doubt and disbelieve, God has given twenty-first-century peoples a tangible, scientifically testable basis for believing that hope, purpose, and destiny are real. Evidence points not only to a reality beyond the cosmos, but to a personal, purposeful, caring Creator and Savior.

Cosmology Offers a Basis for Hope beyond Time

To understand the basis for hope beyond time, one must begin by considering the miraculous timing of astronomers' arrival on the cosmic scene. If they had come much earlier, there would have been less for them to observe, for the objects that comprise the universe have developed and taken shape over the billions of years since the creation event. In an earlier era, there would have been fewer clues to help them discern the wondrous features of the cosmos. The human era just happens to be the best time in cosmic history to be an astronomer.

What is more, the place earth occupies in the heavens also happens to be the optimal location for observing the universe. A team of University of Alabama astronomers led by William Keel have spent more than fifteen years examining the location of our solar system relative to other possible observation sites both within the Milky Way galaxy and without. Their research demonstrates that nearly anywhere else earth might be situated, the view to the galaxy, not to mention the rest of the universe would be seriously blocked.

Where are most of the stars in a galaxy? They are in the globular clusters, the spiral arms, or the central galactic bulge. In all of these locations, astronomical research would be obstructed or prevented altogether by the proximity of stars, including the supergiant variety. Earth's star, the sun, finds itself *between* two spiral arms and far distant from any globular cluster. The fact that the Milky Way holds 150 globular clusters, yet none close to earth, underscores the point. Then there is the fact that earth resides about halfway out from the center of the galaxy—in a zone that makes possible a clear view of the galaxy and of the rest of the universe.

Furthermore, that exact distance from the center enables earth (and its solar system) the rare privilege of *remaining* between two spiral arms. Since the spiral structure of a galaxy rotates at a different rate from that of the stars as they orbit the center of the galaxy, most stars are overtaken and obscured, sooner or later, by the spiral arms. Earth, however, resides at the one distance, called the "corotation distance," where this overtaking does not occur.[20]

This stability of location is critical not just for cosmic observations but also for the possibility of human existence. Earth needs 3.5 billion years of bacterial life—abundant and dominant bacterial life—for subsequent human life to be possible. But, for bacterial life to survive that long, the planet's star must stay between those two spiral arms. If the star goes in and out of the spiral arms, the radiation there and the light from nearby supergiant stars can exterminate the essential bacteria.

Earth's galaxy also enjoys a special location—special with respect to the requirements for human life, that is. The Milky Way belongs to the Local Group. This group includes only about two dozen members (most galaxy clusters contain thousands), and only two are large: the Milky Way and the Andromeda. And the group is well dispersed (in other words, the galaxies are relatively far apart compared to those in other groups). Also, the Local Group resides in the extreme outer fringe of the Virgo supercluster of galaxies. All these particulars of location are significant. Most galaxies, even 13+ billion years after the creation event, remain relatively close together. In most cases neighboring galaxies are so close as to obstruct an observer's view of the night sky. In a galaxy near the center of the Virgo supercluster, an observer there might be able to observe his own galaxy and maybe two or three others, but the window to the rest of the universe would be blocked.

The location, as well as the time, of human existence allows a view to the entirety of the universe out to the theoretical limits. This incredible fact leads one to ask, "Is it just a coincidence, or does it seem that Someone deliberately provided this unique window on the heavens?"

Scripture Reveals God's Purpose for Space and Time

The psalmist offers this answer: God intended that the heavens would declare his glory—*and* much more.[21] For that declaration to be received, its recipients must occupy the just-right time at the just-right place. A powerful statement comes from the fact that we do.

A similar indication emerges from "the anthropic principle"—that the universe seems to be conspicuously designed for human life—a perspective that has been discussed in the science literature since 1961. Princeton's Robert Dicke first made note that certain fundamental forces of physics, in particular gravity and electromagnetism, must be exquisitely fine-tuned for life to be possible.[22] Dicke's work provided the basis for calculating that the ratio of the electromagnetic force to the gravitational force requires fine-tuning to within 1 part in 10 thousand trillion trillion trillion for life to be possible at any time in the history of the universe.[23] The list of finely-tuned, life-essential cosmic features is both long (40+) and growing.

Researchers across a wide spectrum of disciplines and theologies openly acknowledge the validity of the anthropic principle.[24] Those who have done the most research concur that it seems "the universe must in some sense have known that we were coming."[25] Physicist Paul Davies writes, "The evidence for design is overwhelming."[26]

During the 1980s British cosmologist Brandon Carter extended the principle even further, calling his findings "the anthropic principle inequality." Time is the key element. Carter first noted the fact that to get human beings on the cosmic scene in as little as 14 billion years requires virtually perfect orchestration of multiple cosmic factors.[27] Left to ordinary (undirected) physics, life's components would likely take much longer to develop, *if* they ever developed at all.

Three "fortuitous" events in cosmic history gave life a strategic and timely helping hand. First, a type I supernova exploded adjacent to the sun's birthplace (gaseous nebula) 4.5 billion years ago. Second, and almost simultaneously, a type II supernova exploded nearby. Not one but two different supernovae events occurred at just the right time and in just the right location to provide for life's needs—an enrichment of heavy elements. If either supernova had exploded too close to the solar nebula, the forming sun would have been destroyed. But if either had exploded too far away, the nebula would have acquired too few heavy elements to make human life possible in the tight time frame of 4.5 billion years. The word *tight* applies since the kinds, quantities, and convergence of heavy elements produced by these two supernovae would be unexpected in a universe as young as only 9 to 10 billion years (at that time).[28]

Third, a mars-sized body collided with the emerging earth some 4.47 billion years ago. Astronomers now affirm (theoretically and observationally) both the timing and the angle of the collision (neither a glancing blow nor a head-on collision, but something in between). The net result of this event was the removal of the earth's primordial, heavy atmosphere and subsequent replacement with a much thinner atmosphere perfectly suitable for advanced life. In addition, earth was enriched with heavy elements, including uranium and thorium, and a moon (formed from the debris cloud) of sufficient size to stabilize earth's rotation axis for a long time at its (perfect-for-life) 23.5 degrees. Many more benefits accrued from this collision than can be elucidated within the scope of this paper.[29]

Nevertheless, these three amazingly timed and tuned events—some dare say miraculous events—were crucial to the possibility of humans' arrival on the cosmic scene, especially in as short a time as 13–14 billion years.

Carter takes his calculations further yet. He asks, "Once human beings are here, how long can they last in the cosmic environment?" Looking at all the special features life requires of the solar system, the galaxy, and the universe, and at the variability of those factors through time, Carter projected that the window of time for the existence of advanced life is only a few million years wide, at most.[30] Famed physicists Tippler and Barrow, in their seven-hundred-page book on the anthropic principle, propose a recalculation of Carter's figures. Based on various planetary environmental conditions they deduce that the window of time for human survival or, more specifically, for human civilization, cannot be longer than a few tens of thousands of years.[31]

Multiple complex factors affect the brevity of that window, some more restrictive than others. The carbonate-silicate cycle is one factor. This cycle balances the abundances of carbon, sulphur, and carbon dioxide in the environment—a balancing act made remarkably challenging by the increasing brightness of the Sun throughout life's history on earth. The Sun is brighter today by seventeen to eighteen percent than when bacteria first appeared on the earth roughly 3.8 billion years ago.[32] So far, this increase has been balanced by the removal of carbon dioxide, water, and methane from the atmosphere and the conversion of these gases through the agency of life forms into carbonates, sand, coal, natural gas, and oil.[33]

However, as atmospheric carbon dioxide continues to decrease in abundance, green plants will no longer have enough to sustain their photosynthetic reactions. And when the green plants die, everything else dies in succession, with human life among the first to go. Such findings corroborate the narrowness Carter, Tippler, and Barrow have hypothesized.

While Hollywood fuels people's worries about a deadly asteroid striking the earth (an event that recurs every few million years), a more imminent danger lurks in the possibility of a nearby supernova blast. Some 30,000 years ago when the Vela supernova erupted about a thousand light years away, its radiation exterminated several algae species.[34] If, for example, the star Sirius, just eight light years away, were to go supernova, the entire human race would be in "serious" trouble, as one astronomer put it. Not even cockroaches would survive such an event.

As odd as this may seem, human extinction may be hastened most significantly by such mixed blessings as technology and affluence. The more advanced and affluent a society becomes the fewer children that society produces and the longer people postpone having children. The longer people wait to have children, the more negative mutations they (the fathers, especially) pass on to the next generation.

In the last hundred years, growing affluence and technology have resulted in men's having children about seven or eight years later, on the average, than they did 150 years ago. In this period, the negative mutation rate has escalated dramatically. Attempts to measure this increase place it at three negative mutations per person per generation.[35] Extrapolating into the future, one sees a bleak forecast for the human species' survivability.

Affluence and technology also allow individuals who would have died in childhood, under former circumstances, to survive into their reproductive years. Thus, individuals with high numbers of deleterious mutations have greater opportunity to pass those mutations along. In a recent private conversation, economist Michael Phillips speculated, based on "the affluence problem" alone, that earth's population could drop to as few as 1 billion people by the year 2100. His back-of-the-envelope forecast is based on income and birthrate statistics. In nations where average (adult) annual income exceeds $20,000 per person, the reproductive rate is only 0.7 children per adult. Again, extrapolation makes the problem apparent.

The key point Carter seeks to drive home, a point with which my reason concurs, is this: If it takes nearly 14 billion years (at a minimum) to prepare a home for humanity, and if the window for human existence is only a few tens of thousands or even a few millions of years wide, one cannot escape the impression that the human species carries a high value.

This impression emerges as an echo of ordinary human experience. People tend to invest most lavishly on the individuals and relationships they value most. In my own case, for example, my love for my sons motivates me, as thrifty as my Scottish heritage makes me, to spend time, money, and other resources on what means a great deal to them—even if I know their interests and needs may change in a few years.

The extreme inequality between the time required to provide a home for humanity and the brevity of humanity's existence plausibly suggests that the Creator of the universe and of human life intended that human beings come into existence *and* that he cares a great deal for them.

Unique drives of humans seem to corroborate that sense of "intended" existence. In addition to the powerful survival instinct they share with all sentient life, people manifest a unique drive to discover and fulfill their destiny. A sense of hope enables them to survive even horrific circumstances, and a sense of purpose enables them to thrive. For some people, these drives intensify with age, but in others they are evident from youth. At almost any age, a person may be driven more powerfully by this sense of purpose than even by the instinct to survive. Familiar examples would be Dietrich Bonhoeffer and Mother Theresa, whose devotion to others superceded their instinct for self-preservation.

Where do these drives come from? Researchers find no evidence of them in humans' supposed ancestors, the primate species, including the bipedal hominids. People's otherwise inexplicable, often self-sacrificial yearnings make sense only in the context of creation, specifically in the context of a personal Creator's plans, purposes, and participation on behalf of human life.

My studies in physics provide me with yet another compelling evidence for the reality of the loving Creator and of human destiny both *in* and *beyond* cosmic time. That evidence may be described as the optimization of physical constants and laws to bring about the ultimate expression of God's love, which involves allowing and then conquering the possibility of evil.

A close and careful look at the gross features of the universe, at the characteristics of space and time, at the laws of thermodynamics and at the constants of physics, reveals that all were designed and implemented for the physical and spiritual benefit of humanity—given the inevitability of original sin.[36]

When Adam and Eve rebelled against God's authority (Gen 3) and introduced evil to the human race, God explained to them what the consequences, or "curse," would be. One could argue that he simultaneously preserved the hope of humanity.

Sin's consequences were, among other things, troublesome work and grievous pain.[37] In some sense, the work and pain humanity suffers is generally proportional to the expression of spiritual autonomy, or rebellion. God commanded the first humans (Gen 2:15)—and all humanity through them—to "tend" not just the garden he planted, but also to manage wisely the resources of the whole wide, *wild* earth (Gen 1:28-30). In humans' relationships with each other and with the environment, more sin means more suffering. The more the human race expresses pride, selfishness, and greed, and the less it depends upon God for humility, love, and wisdom, the more damage it inflicts. The damage to society and to the planet results in suffering of all kinds. The desire to avoid troublesome work and grievous pain provides some motivation to refrain from sin and evil.

Given the realities of human (sin-marred) nature, one begins to see the temporal and spatial limits of the cosmos as divine blessings. Because time, as humans experience it, is one-dimensional, irreversible, and unstoppable, time imposes a limit on the quantity and degree of evil any human can perpetrate against others. Though time may also limit the quantity of good a person can do, the early era of human history, during which lifespans were considerably longer, indicates which direction—toward good or evil—the balance tips. Extended lifespans led humanity to the brink of extinction, to the necessity of a cleansing cataclysm, the Genesis flood.

More recent examples of this point appear glaringly in the lives of such despots as Adolf Hitler, Joseph Stalin, Mao Tse-tung, and Saddam Hussein, to mention just a few. Apart from the limits of time—and space—these individuals would have wreaked even more havoc and harm than they did. Each had known plans to do so.

This observation would seem to turn us back to the dark despair articulated by Krauss and Starkman. It leads one to ask, "Will these limits always be necessary?"

CONCLUSION:
SCRIPTURE REVEALS GOD'S PLANS FOR TEMPORALITY BEYOND TIME

Science attests what Scripture reveals—that something or, more accurately, Someone, really does exist beyond matter, energy, space, and time. Hope, purpose, and human destiny are all connected to that greater reality. The Bible presents a "two-creation" reality, a conception utterly unique to Christianity. While other religious systems may promise some kind of heaven or alternate reality, that "paradise" falls pitifully short of what Scripture describes as "the new creation." Even what Adam and Eve experienced in the garden of Eden before they sinned falls short of what God has in store for his people.

Rather than promising the restoration of paradise, or Eden, the Bible promises *deliverance from* an earthly home, however wonderful it may seem. If one looks at space, time, and physics in the garden of Eden, one sees the confinement of humanity, a set of limitations that is removed in the new creation, just as the Tree of Life was removed from the old.

The two-creation scenario offers a refreshing vantage point from which to view the past, present, and future of God's relationship with humanity. The Bible reveals a universe and earth meticulously prepared for life, and for human life in particular, over the course of its long history. When the preparations were complete, God created Adam and placed him in a paradisaical garden, instructed him concerning his responsibilities, and created a helper for him. Then God allowed Satan to enter that garden. The sovereign, omniscient, omnipotent God could have barred Satan from entering, but he did not—though he knew exactly what would happen.

God knew the humans would be enticed by evil, that they would fail the obedience test. But his plan was in place. Through the power of the incarnate Son, Jesus Christ, the "second Adam," he atoned for sin, breaking through the boundaries of matter, energy, space, and time. The Holy Spirit began a new work in humans (those who accept God's gift of redemption), to deliver them from the grip of evil and move them toward the full-blown expression of his power and love.

Even now the new creation is being prepared for those humans who, as the Spirit enables them, exchange their propensity to sin and even their best efforts against sin, for God's gift of righteousness by faith. When time's purpose is fulfilled, when time has allowed all the people of God's choosing to choose him, the court of the Great White Throne will convene. In that moment of final judgment, God permanently binds up evil, separates it from his people, and brings them into his presence—into what he describes as an entirely new creation.

The preview he gives shows how new and different it must be. Its physics are different. Its dimensionality is different. Its "geography" is different. Its "temporality" is different. God makes it different to set redeemed humans free from cosmic relational limits. God will be with us and we with him in a completely new way that exceeds, as the apostle Paul says, what anyone can even "think or imagine" (1 Cor 2:9).

In other words, no human mind can fully appreciate what the second creation will be. The first creation is perfectly designed to prepare us, through God's redeeming work, for the creation that is yet to come. The new creation allows us to live beyond cosmic limits in the presence of God. The new creation holds incomparable splendor, joy, beauty, love, and light. With a view to or from this future, our space-and-time-bound existence will seem—no longer paradoxically—infinitesimally brief and yet eternally significant.

C. S. Lewis offers this glimpse from the last page of the last book in the Narnia Chronicles: "And for us this is the end of all the stories. . . . But for them it was only the beginning of the real story. All their life in this world . . . had only been the cover and the title page: now at last they were beginning Chapter One of the Great Story, which no one on earth has read: which goes on for ever: in which every chapter is better than the one before."[38]

Chapter 8

Meeting the Cosmic God in the Existential Now

TONY CAMPOLO

C. S. Lewis was well aware that modernity had its dangers. In *That Hideous Strength*, and *The Abolition of Man*, he cited what those dangers are. He saw that science, technology, and reason, with the innovations they fostered, were not producing the progress that everyone assumed that they would produce. There was supposed to have been a Utopia that would emerge out of positivism and scientific investigation, but it just was not happening. Quite the opposite was becoming increasingly evident. Science was turning into a Frankenstein's monster that threatened our existence. We live with the fears of what cloning might usher in and we are beginning to be afraid of what computers might do to us. Right now at Harvard University, scientists involved in artificial intelligence experiments predict that computers will transcend human intelligence within ten years. What is more, they expect that computers will be able to replicate the brain and actually stimulate a part of that replicated brain so that it can have religious experiences.[1] This is something we will have to deal with as the whole new field of neurotheology emerges on the scene. These were things that C. S. Lewis saw coming, and he stood against them all.

In spite of his fears about threats that were posed by our increasingly technological society, Lewis saw that science, in its purest sense, pointed beyond itself to something spiritual. He was no enemy of science. He did not, as some postmoderns do, discard science and read it off the charts. He saw that it had potentiality for nurturing godly thinking. Science, he saw, could point beyond itself. He watched with great interest the findings of science especially the cosmology being developed right there at his beloved Oxford University. He was very aware of what was going on in the laboratories and observatories of his scientific colleagues. Lewis was a contemporary of Sir Arthur Eddington, the astronomer laureate of Britain, who wrote *The Expanding Universe*.[2] He was in dialogue with Samuel Alexander, whose book *Space, Time, and Deity* influenced thinking.[3] He was aware of theories being developed by Einstein. He understood many of the implications of the new cosmology that were shaking the foundations of the established *Weltenschauung* of the day. As a young graduate student, I personally found this new cosmology that followed Einstein's discoveries extremely useful in dealing with the intellectual challenges being presented in the halls of academia, and I cannot help but believe that Lewis did the same, to an even greater degree.

When I was coming through school, the ideas of Rudolph Bultmann were all the rage. Bultmann had taken the existentialist philosophy of Heidegger and baptized it to make a new theology. I remember that Bultmann's argument was that the New Testament was written against the backdrop of an ancient cosmology that was no longer operative.[4] As Bultmann looked at history, he said there was an ancient time when people believed that they lived in a three-storied universe, and that the biblical message only made sense against that backdrop. The three-storied universe with heaven above, hell below, and earth in between seemed real to the biblical writers. They believed in a heaven "up there" from which angels descended and to which angels ascended. They believed in a hell under the earth. Such a universe was the context in which the biblical stories made sense. But all of that has changed now, said Bultmann, and the cosmology that has taken its place through Newton's discoveries, he claimed, has made it impossible for us to adhere to a theology that is wedded to this three-storied universe.

"The universe is infinite," Newton said. It goes on and on and on, forever and ever and ever. In such a universe, according to Bultmann, the Scriptures make no sense. In this new cosmology there is no longer an "out there." There is no "up there." With the new cosmology, contended Bultmann, there is no place for a heaven over a hell. These places that were spatially located in the Bible have no place in the universe that emerged from Newton's physics. The intriguing thing was that what Bultmann said all sounded so good and made a great impact on impressionable minds like mine in the early 1950s, but of course he was a little behind the time. Newton had already been challenged by Einstein.

Einstein came up with that famous theory of gravity, and early on scientists realized that this new understanding would not work in an infinite universe. Lesser minds would have concluded, "Well, if my theory does not work in an infinite universe, then I guess I must discard it." Not Einstein! Still in his early twenties, he had the arrogance that people have at that age. He said that if his theory did not work in an infinite universe, it was because the universe is not infinite. Imagine that: "If the universe does not agree with me, then our understanding of the universe is wrong!"

It was Alexander Friedman, a colleague of Einstein's, who saw a resolution to the problem. Friedman was the first really to begin to define the universe as limited but unbounded, and Sir Arthur Eddington at Oxford carried it even one step further. What these scientists postulated, which empirical research has now validated, is the fact that the universe is actually unbounded, but limited. It is all beyond me. According to, Eddington the universe had resulted from the "Big Bang," about 15 billion years ago. That was when an infinitely small photon suddenly exploded, and the universe as we know it is the result of that explosion. Time and space and matter have all emerged out of that Big Bang. Once, when I personally heard Einstein explain this theory, he said that we must compare the universe to a balloon that is in the process of being blown up. Time, space, and matter are, he said, all in the skin of the balloon and, like the universe, are expanding outward. If you put dots on the skin of this balloon then, as the balloon expands, these dots appear to move further and further apart from each other. These dots, explained Einstein, represent galaxies, and, like the universe, as this balloon expands, these galaxies likewise appear to be moving further and further apart. What Einstein said that day has been validated by findings made by the Hubble telescope.

It was Sir Arthur Eddington who asked the obvious follow-up question. If the universe is likened to an expanding balloon, or an expanding puff of smoke, one may ask, "What is the nature of the larger reality in which this expansion is taking place?"[5] What is the realm beyond which time and space and matter exist? In short, all of a sudden it becomes logical and reasonable to talk about transcendence once again. If time and space and matter expand outward, is it not reasonable to assume that there is a larger arena—something beyond time and space and matter? That is an intriguing question, to say the least.

Time as a Dimension

Another corollary of this new cosmology was that there was a fourth dimension to the universe, and this fourth dimension left room for the seemingly miraculous to happen. Lewis had long believed in miracles, and I remember what he had to say about them.[6] His wonderful illustration in his book on miracles made everything seem so clear. For instance, he pointed out that some primitive people living on an obscure island might believe that anything heavier than air cannot fly. Their empirical testing had proved that. They pick up rocks, let them go, and they fall. Anything heavier than air falls, they conclude.

Then one day this metal thing with shining wings flies overhead. It is an airplane. The airplane is heavier than air, but it is flying! To the people on this island, it is a miracle, but in reality, what is happening is that those who designed the airplane understood more about science than did the people on the island. They understood that there were certain laws of thermodynamics that are such that when an object heavier than air, constructed in a particular shape, goes at a particular speed, it can fly. What the inhabitants of the island called a miracle, those with more advanced scientific knowledge understood was not really a miracle so much as it was taking advantage of scientific realities that were beyond the comprehension of the people on that little island.

For C. S. Lewis there were not two kinds of knowledge. He reacted against the German schools that had *Geisteswissenschaften* knowledge on the one hand,

and *Naturwissenschaften* knowledge on the other. This division posed that the natural sciences had one kind of truth and those in subjects such as religion had another kind of truth. This view argues for two kinds of truth. C. S. Lewis would not accept this divide. For him, truth was one. For him, all things had to be integrated into one unified system.

Lewis would have gotten along well with Einstein, who also believed that all truth could be integrated into a single system, even though he did not know what the formula for that system was. Einstein also believed that what we consider miracles were really scientific possibilities, especially within the new worldview he was helping to develop. To show how new apprehensions of reality can make miracles possible, he often had his students read *Flatland*. This is a whimsical little story about a man from a three-dimensional world, our world, who suddenly ends up in a two-dimensional world.[7] Everything in Flatland is flat, and in the world of Flatland, the three-dimensional man is able to perform miracles. In Flatland, a line is an insurmountable barrier to the natives of Flatland. They cannot go through it. But ah! Our hero is three-dimensional. He is able to step over the line. Everybody in the two-dimensional world says, "Miraculous!" In reality, this visitor to Flatland is only taking advantage of another spatial dimension the others cannot comprehend. Another miracle he can perform is to disappear. In Flatland, he is able to disappear simply by jumping up, because in Flatland there is no "up." Everything is flat. He can disappear from sight instantaneously. All the people in Flatland say, "Miraculous!" In reality he is only taking advantage of a third dimension of space.

What Einstein's theory let loose on the world was the idea that the world was not actually three-dimensional. Instead, it was four-dimensional. What is more, Max Planck and his followers calculated that there might be up to eleven dimensions in the universe. Think of the implications of all of this for a resurrected Jesus who steps into our time and space, having once again taken on the capacity to utilize the possibilities of a multidimensional universe. Might such a Christ be able to take advantage of multiple dimensions of space? Might such a resurrected Christ be able to pass through shut doors (John 20:19) or instantaneously disappear (Luke 24:31)?

Time: Biblical and Earthly

Oscar Cullmann, noted for his biblical analysis of how time is regarded in the Bible, contends that time in Scripture is defined as a linear progression reaching from the infinite past, prior to creation, and extending into the infinite future. Historical time, according to Cullmann, is simply a segment out of the linear development that he believes to be an everlasting progression.

On the contrary, Emil Brunner, the Swiss-German theologian, argued that historical time was a created realm within the larger reality of "God's time," which Brunner called "the eternal now." In this eternal now, said Brunner, time as we understand it is transcended, and the concepts of *before* and *after* that go with our linear concept of time become meaningless.

Such arguments about the nature of time and its relationship to what the Bible defines as eternity, are nothing new to theologians. Theologians have been intrigued by discussions on the subject from as far back as Saint Augustine. It was Augustine who, early in the life of the church, recognized that a host of theological questions could be resolved if it were possible to figure out how time as we know it is related to God's time. Augustine, centuries before Einstein and somewhat anticipating Einstein's thinking, taught that time was related to creation, and that it had a beginning when God spoke the universe into existence. He then raised the paradoxical question as to whether or not there was a time before there was time. As we shall see, an answer to this question is full of theological intrigue.

Surprisingly, Augustine's problems regarding theological implications for time might find some resolution in Einstein's physics. First of all, let us consider what Einstein found out about time. He came up with the amazing discovery that time becomes compressed with increased speed. In other words, the faster we travel, the more time is compressed. While the following figures are not precise, they may help in understanding what this is all about.

+ If you were put into a rocket and sent into space, traveling at 170,000 miles per second relative to earth, and then returned in ten years, when you returned you would be ten years older, but all the rest of us would be twenty years older. For those of us who remained on earth our twenty years would be compressed into ten years of your time.

- If we got you traveling at 180,000 miles per second, our twenty years would be compressed into one day of your time.

- If we could get you traveling at the speed of light—186,000 miles per second—there would be no passage of time at all. If that were possible, all of time would be compressed into one eternal now.

Young people who have watched the television show *Star Trek* have no problem with any of this, because they are used to Captain Kirk and company moving forward in time simply by altering the speed at which the spaceship *Enterprise* travels.

Time and "The Now"

In reflecting upon how time is relative to motion, I want to declare that I believe God is able to experience time at the speed of light. For God, all of time can be compressed into what the theologian Emil Brunner called "the eternal now." All things happen "now" with God. With God, a thousand years are as a day, and a day is as a thousand years (Ps 90:4). The very name of God implies this reality. God's name is "*I Am*." God never "was," and God never "will be." God is always "now."

The "nowness" of God lends support to the declaration of the theologian Paul Tillich, who contended that God could not be known as an object caught in the space-time continuum in which we live. According to Tillich, God can only be "encountered in the now." We cannot analyze God and then describe God in ways that fit the categories of time-bound logic. God will not yield to positivism (the philosophy that reduces all reality to terms that the five senses can experience), nor be understood in terms of the categories of human logic.

Those philosophers and social scientists who call themselves *phenomenologists* are most in harmony with the concept of the eternal now. They point out that "now" has no extension in time. They argue that we cannot say that the next minute is now; nor can we even say the next second is now. We cannot even say that the next millionth of a second is now. "Now," they contend, is that nonexistent point wherein the past meets the future. "Now," in a sense, is not really in time. Yet each of us can say, "I am alive now! Now is where I am!"

Everything that is true about *me* belongs in the past. I can reflect on me—I can take a good look at myself—but everything that I observe about me is in the past. What I can know about me is not who and what I am right now. "Me" is what I used to be. I cannot even tell you what I am thinking now, because in order to do that, I have to stop and consider what I *was* thinking. But again I point out that now is where I am—it is where I exist.

More importantly, God is in the now. For God, all of time and eternity are gathered up in this eternal now. God is the Alpha and Omega; the beginning and the end (Rev 1:8). Given this assertion about God's being in the now, we can begin to understand why Tillich argued that God cannot be known as other objects are known, because in the now, there is no objective knowledge. That is because objective knowledge requires reflection, and when we reflect, we can only reflect on what was—never on what is now. If I ask you to reflect on what you are experiencing now, you can only tell me about what you were reflecting upon when I asked you the question.

God in the Now

All of this seeming double-talk affirms a basic evangelical truth, and that is that those who would know God must encounter God in the now. Right now, the person who would know God must surrender to God, and let God overwhelm and invade his or her mind, body, and soul. A person does not become a Christian simply by gathering objective knowledge about God. In reality, an individual may know all the doctrines about God and have a "sound" theology about what God accomplished in the death and resurrection of Jesus Christ, but he still may not *know* God. There is a qualitative difference between knowing all about God and having a sacred encounter with God in the now.

Each morning I try to wake up half an hour before I have to, and then lie in bed, experiencing quietude. In the stillness of the morning, I surrender to a possible encounter with God. I wait patiently to feel God's presence. I sense a beckoning, as did the psalmist of old, who heard the admonition, "Be still, and know that I am God" (Ps 46:10). The "still small voice" of God has no words (1 Kgs 19:12), but there comes an inner groaning as I encounter God in the now.

I can talk about this experience with God. I can try to describe it and even theologize about it. But in doing so, I am no longer in the experience. I am only reflecting on the experience that I just had with God. In such occasions of mystical ecstasy, time, in the linear sense, ceases to exist. I am, in such spiritual encounters, beyond time. I am experiencing God in the now. This is the kind of knowledge about God that the apostle Paul was praying to experience when he wrote, " I want to know Christ and the power of his resurrection and the sharing of his sufferings by becoming like him in his death" (Phil 3:10).

Such an encounter with God is one way to experience what we evangelicals call being "born again." It taps into the essence of the conversion experience.

Einstein's Theory and the Cross

A great and perhaps most important insight that can come out of the relativity theory is a deeper understanding of what Jesus accomplished for us on the cross. If you believe, as I do, that Jesus is fully God as well as fully human, then you might be ready to accept that there are some existential dimensions of his saving ministry on the cross that Jesus is doing now. As a man, and the only sinless man who ever lived, Jesus lived and died some two thousand years ago. Back there and then on the cross, he took the punishment for our sins. Jesus was the sacrificial Lamb of God—the sacrifice that delivers each of us from the condemnation that is our due because of the dark side of our humanity.

When he died there on that hill outside of Jerusalem, Jesus became the substitute who went through the hell that should have been our lot because of the dirt and ugliness of our sins. That he endured this punishment on our behalf is a historical fact, established in linear time. It is a done deal! There is nothing more to be done to deliver us from the punishment that should be meted out to each of us. We can justifiably talk about the finished work of Jesus. Historically, in time, the perfect sacrifice of his life for ours is an accomplished fact.

Something More

What I have just stated relates to what Jesus did for us back there and then. But there is something more that the would-be Christian should grasp to experience all that Christ's salvation has to offer. In addition to what Jesus did for us, each of us must be ready to surrender to what Jesus wants to do *to* us in the *now*. Because Jesus is not only "fully human" but also "fully God"—even as the Chalcedonian Creed declares—he is able to comprehend time as God. Because he is God, Jesus is able to gather all of time together in his eternal now.

Jesus not only was a man who lived, died, and was resurrected historically; he is also the God for whom all things are in the present moment. That is why he could say to the Pharisees, "Before Abraham was, I am" (John 8:58). Jesus wasn't using poor grammar when he said that. Instead, he was declaring his divinity. He was saying that the time before there was an Abraham is present tense for him. Because of his divinity, he experiences all events in time and history as happening for him in the now.

If you are following all of this, you may be ready to grasp what for me has been one of the most important dimensions of what Christ did and does on the cross. When he hung on Calvary's tree some two thousand years ago, being God, he was—and he is—simultaneous with me here and now. Right now, I am caught up in his eternal now. The centuries that separate me from Jesus suffering on Golgotha are compressed, as though at the speed of light, so that for him, he is with me in this instantaneous now. That means that *right now*, in God's divine sense of time, as he hangs on the cross, he is able to empathize with me, and via a kind of spiritual osmosis, absorb into himself all the sin and darkness of my life.

On the cross two thousand years ago, he took the punishment for my sin, but right now, in his eternal now, he is able to reach out to me from that old rugged cross. Like a magnet, he can draw out of me all the evil that is part of my humanity, as though it were some iron filings.

The Bible says: "For he hath made him to be sin for us, who knew no sin; that we might be made the righteousness of God in him" (2 Cor 5:21 AV). Here and now, if I surrender to him, he will purge me and make me pure. If he is in me, and I am in him in his eternal now, then I become a new creation. Old things pass away, and everything becomes new (2 Cor 5:17). "Though your sins be as scarlet, they shall be as white as snow; though they be red like crimson, they shall be as wool" (Isa 1:18 AV).

There is more to salvation than just being delivered from punishment for sin. There is a *cleansing* of the heart, mind, and soul that can occur in the now. It was for this special existential cleansing that King David prayed in Psalm 51. And it is this dimension of Christ's work on the cross that we read about in 1 John 1:19, "If we confess our sins, he who is faithful and just will forgive us our sins and cleanse us from all unrighteousness."

In this verse, we not only read about the forgiveness of sins that comes from what he did back on Calvary, but we also learn about the cleansing that can occur in the here and now. Jesus, being God, can cleanse us today because he is, in the words of the Danish existentialist theologian, Søren Kierkegaard, "the eternally crucified." When I preach evangelistic sermons within the church, I usually say, "I know you believe in what Jesus did for you back there and then. But what I want to know is whether or not you are willing to let Jesus do something to you *right now*."

I want to know if you are willing to let Jesus reach out to you from the cross and absorb your sin, making it his own, and at that same instant let his goodness flow into you to take the place of sin. I ask, "Will you let him impart to you his righteousness? If you will, then on that day of judgment, the one who has taken your sin and made it his own in the now will present you to his Father, as it says in the book of Jude, *faultless!*"

Motivation for Holiness

I do not want to leave the discussion of what goes on with Christ's suffering and cleansing us from our sin in his eternal now without pointing out that all of this should provide a great impetus for living the holy life. Every time I sin, that very instant Jesus groans in agony in Calvary. Even as I sin today, he experiences the agony of ingesting my sin into himself in his eternal now, as he hangs "spread eagle" on the tree back there and then. That is why it says in Hebrews 6:6 that when we sin, we crucify him anew. In a sense, when we sin, we crucify him right now.

I was at a Christian college that is one of our citadels of evangelical purity, talking to a student who was a bit troubled over the fact that he was an evangelical Christian, but he was having sex with two different girls. I said, "Well, how do you reconcile all this with being a Christian?" He said, "Well, I believe that Jesus took care of it all long ago and far away."

I said, "The next time you're committing fornication I hope you can hear Jesus screaming in pain in the background. Because at that very moment he is simultaneous with you, absorbing the sin that you are committing into his own body." No wonder Paul says that we dare not sin that grace may abound (Rom 6:1) because even as we sin, two thousand years ago on the cross, Jesus reaches across time and space, and draws those sins into his own perfect body, as though they were iron filings and he a magnet.

First John 1:9 tells us this: If we will confess our sins he is faithful, he is just. He will forgive us of our sins and he will cleanse us. Reformed theology bases that forgiveness on what he did back there and then on Calvary. But something has to happen right *now*! You have to surrender in stillness to Christ on the cross and let him *cleanse* you. And do not tell me you do not need cleansing.

Toward a Resolution

Of course, the resolution of the death of Christ and the second coming of Christ is resolved. I have some Seventh Day Adventist brothers and sisters whose theology says that when you die you lie in the grave until the resurrection morning. I think that view is sustained in Scripture, so that affects the way in which you should be buried. I want to be buried with one hand up, waving!

Then there are those who would maintain that the moment you die you are simultaneously present with Christ, and I think that position can be legitimated scripturally. How can these two seemingly contradictory things be true? How can the Adventists be right, and the Baptists be right at the same time? Well, in linear time, those events are separate. If I were to die today and the second coming was to occur (and I don't know when it will occur—I mean, I'm on the welcoming committee, not the program committee) ten years from now, I would be in the grave for ten years—agreed? The body is going to be resurrected. The Scriptures do not teach the immortality of the soul, strange as it may sound. The Scriptures teach the resurrection of the body. My body will not be resurrected until the resurrection morning. Ah! But the moment of death and the moment of the second coming at the speed of light are what? The same moment. I hate to tell you Adventists—we are both right about when both events occur. From a historical linear perspective where we live in time and space, they are separate. Of course, I will not even begin to suggest what the implications of this are for people like Ben Patterson who are Calvinists and believe in predestination.[8] Does God know what is going to happen before it happens?

The question is a simple one. Does God know before? Well, what if there is no "before" with God? Suppose he is the Alpha, the Omega, the beginning and the end. Suppose everything is "now" for God. The intriguing, subtle implications of new time theory emerging out of relativity are startling, to say the least.

Go a step further. C. S. Lewis saw this. As he reflected on the concept of prayer he saw a problem.[9] He wondered how God could give himself totally to any given individual at any given moment. "How can God be totally committed to me when I pray," he asked, "and simultaneously be wholly committed to every other individual in the world who might be praying at that same moment?" As Lewis began to reflect on the new concepts of time that were emerging, he realized that if at any given moment there is an infinite amount of linear time, since all of time is compressed into the now at the speed of light, then it is possible that at any given moment God would have an infinite amount of linear time in which to hear the prayers of each and every person who is praying at that moment!

When you realize where C. S. Lewis was when he was writing, you must also realize that Einstein was still off on the side edges of intellectual thought. Relativity theory had not yet been popularized as it is today. Now everybody knows something about relativity theory. We watch Star Trek, and we know we can go forward in time just by changing the speed of the Starship Enterprise. You knew that, didn't you? And if you saw the movie Contact, you know that Jodie Foster was afraid to get in the space vehicle and go to another galaxy and come back, because she would be traveling so fast that when she returned her lover would either be dead or so old that it would not make any difference.

It is amazing to me that the more science expands, the closer it comes into agreement with the teachings of God's word. C. S. Lewis believed that. So, to a Bultmann, who throws the Bible out because he does not adhere to a three-storied view of the universe, I say, "Look again. You're a very clever fellow, Bultmann, but you're behind the times."

Of course, the modern cosmology, as C. S. Lewis saw, postulates that time had a beginning. There was a time, according to the Big Bang theory, when it all started, some 15 billion years ago. That is not so long ago in the context of eternity. Fifteen billion years ago a photon exploded and time was one of the results of that explosion. There was a beginning to time. Recognize the theological implications of that beginning.

One of my friends in the field of social history contends that some very important words in the Bible were the first three words: "In the beginning . . ." Please understand that in some other religions there is no beginning. In Buddhism and Hinduism, time moves in a cycle. It is an endless cycle. As a matter of fact, the book of Ecclesiastes gives us a glimpse of what results from a cyclical view of time. The grass grows, the grass dies, it grows up again, and it all repeats itself over and over again.[10] Friedrich Nietzsche embraced this concept of time in his theory of recurrence.[11] Certainly those who are into the New Age movement are very much into a recurrence theory. It all repeats itself over and over and over and over and over again. To those who embrace a recurrence theory, there is no beginning; there is no end. Hence, there is no meaning to life. Instead, as the author of Ecclesiastes states, "All is vanity."

With Big Bang cosmology, however, the recurrence theory becomes untenable. There was a beginning. The one law of thermodynamics that continues to exists in our post-Einsteinean era, is that energy is being dissipated in an exploding and expanding universe. The universe is moving to maximum entropy. Consequently, there will be a time, even as the Bible says, when time shall be no more. If time is relative to motion, when all energy is dissipated, and all motion ceases, time will cease to exist. No wonder Stephen Hawking, who tends toward atheism, came up with a recurrence theory in his cosmological theories.[12] Hawking is smart. He knows that the implications of a Big Bang theory without recurrence will push towards theism, and he is afraid of that. Thus, he proposed a theoretical framework which I do not think can be empirically verified. The expanding universe has been empirically verified, but I do not know how you could verify the recurrence theory.

There are atheists who have begun to develop a whole new field of theoretical reflection called Precosmic thought. They know that if there is no God, they will have a hard time explaining how the universe happened. The Big Bang theory continues to haunt those cosmologists who do not have any grasp of theism. Einstein's last book, interestingly enough, is entitled *Metaphysics*, for he saw that there were implications—metaphysical implications—to what he was all about. The universe has a beginning, and it is moving towards an end.

Lewis was a contemporary with Teilhard de Chardin, and like Teilhard, Lewis saw that history had a purpose to it. It is moving toward some kind of fulfilling climax. Those who have read *The Phenomenon of Man* know this: Teilhard saw that there was an emergent quality to the universe, with galaxies being formed through a process which he called "amorization." Through this process he saw the love of God at work in the universe, bringing cosmic dust together to form galaxies. God is viewed by Teilhard as the great unifier, bringing (together) nonorganic matter in such a way as to create what he called the biosphere, that is, the biological world. Out of the biosphere, Teilhard claims, comes the noosphere—the human mind in *homo sapiens*. But Teilhard contends that there is still a future stage for the cosmos. He calls it the Christosphere. In it, human beings come together and unify, forming what the Bible calls "the body of Christ." This is the eschaton. This is when all humanity is harmonized into a loving entity.[13] What I wish he had said was that there is something coming called the kingdom of God.

We believe that there is a logic and reason in history; that the God who set the universe in motion has a plan, and that the plan is the actualization of the kingdom of God within the cosmos. The kingdom of God is no pie-in-the-sky-when-you-die-by-and-by sort of thing. It happens in history. When I pray the Lord's Prayer, I pray this: "Thy kingdom come, Thy will be done *on earth* . . ." The kingdom of God is where sociology and cosmology connect, because in the context of this new cosmological construct, I believe that something is happening in history. I do have a view of progress. I believe that God is at work in the world creating his kingdom. T. S. Eliot was brilliant with words, but he was wrong when he wrote that couplet: "This is the way the world will end, not with a bang but a whimper."[14] Then with dissipation add, "Ah, man . . . ."

Eliot was wrong. It will not end with a bang, and it will not end with a whimper. This is the way the world will end: the kingdoms of this world will become the kingdom of our God, and he shall reign forever and ever (Rev 11:15). Hallelujah, hallelujah! That is the view of history that I preach! There is a *telos* (goal) to history. The universe is going somewhere; history is going somewhere; it is moving toward the realization of the kingdom, and we are called by God to join him in making this happen.

In the prayer this morning, Nigel Goodwin said, ". . . that we might work to build the Kingdom."[15] That is not some spiritual abstraction. The kingdom is a concrete sociological reality. In Isaiah 65:17, we get a description of that kingdom. Isaiah says that in the kingdom of God, everyone will have a decent house to live in. Everyone will have a job. We will be blessed by the labors of our hands in the vineyard. Nobody else will steal the fruits of production.

In the kingdom of God, children will not die in infancy. Look at Africa today with the AIDS crisis and the many children who are dying there in infancy. It will not happen in the kingdom of God.

In that kingdom, old people will live out their lives in perfect health. Isaiah says that the man who dies at a hundred will be considered to have died young. We already know that with the new genome studies they have already isolated the gene that controls aging. They have already been able to deal with flies and mosquitoes and get them to live ten times their normal life span. (I do not see the point of *that*!) But by the grace of God, humanity will be able to expand life expectancy. We will be able to engineer certain diseases out of existence, and by the grace of God people will live longer lives in good health.

Isaiah says that when children are born, they will not be born to calamity. How many children are born in the ghettos of Philadelphia, in the slums of Los Angeles, and in the hard places like Brixton, where the parents at the day of birth wonder whether this child is being born for calamity? Will this child be freaked out on drugs? Will this child be blown away in a gang struggle? Will this child end up pregnant before her time? What will happen to this child? No more will they be born to calamity (Isa 65:23).

For the environmentalists who are here listen to this: neither shall they hurt the earth any more (Isa 65:25). What an image! Environmental responsibility. What an image! Infant mortality obliterated. Decent housing. Good jobs.

The kingdom of God is not an abstraction. It is a historical reality. We are moving towards it. Sometimes we get discouraged, but in the midst of our discouragement we must remember World War II. When we recognize how history will end, we know that we are on the winning side. Oscar Cullmann, in his book *Christ and Time*, points out that in every war there is a decisive battle that determined the outcome of the war.[16] In the Civil War in the United States, it was Gettysburg. After Gettysburg, everyone knew how it was going to end. The war went on after that, but there was no question about how it was going to end. In the Napoleonic wars it was Waterloo. After Waterloo, everyone knew how it was going to end.

In World War II, the decisive day was D-Day. The Nazis knew if they could drive the Allied forces back into the sea they would win the war. On the other hand, there was a realization that if the Allies could establish the beach-head, victory for the Allies was inevitable. It was called by many, "The Longest Day," because history hung in the balance.

The good news is that at the end of the day, the beachhead was established. After D-Day, there was never any question as to how the war was going to end. No wonder Rommel joined the plot to assassinate Hitler. He knew that Hitler would never give up, and at D-Day the war had been lost.

There would still be a lot of death, a lot of suffering and a lot of agony after D-Day. More people would die after D-Day than before D-Day. Yet in spite of the suffering and the death and the pain, there was never any question as to how it was going to end. V-Day was inevitable! Oscar Cullmann pointed out that we now live between God's D-Day and God's V-Day. D-Day was the death and the resurrection of Christ, where all the forces of the demonic were mustered and focused on one man nailed to a tree. If ever Satan was going to win the victory, it would be there. It would be then. At Calvary, God was incarnated into the most vulnerable of forms, into the weakness of human flesh, says the Scripture, emptied of power and hanging spread-eagled. What could be more vulnerable? When the day was over, it looked as though Satan had won and the dark forces had won. It looked as though the forces of darkness would rule the universe. But three days later, Jesus staged a coup. Up from the grave he arose! *Christos Victor!* That was God's D-Day! But the war is not yet over. The decisive battle has been won, but V-Day is yet to come. I do not know when it will happen, but it will happen. From now to the end, we carry on in the struggle, but we know that in spite of our weaknesses and shortcomings, *we will win!* That is because we believe in Christ's return.

If you had asked the French Underground, "Do you really expect to defeat the Nazis with the small group of soldiers you have and a few machine guns and a handful of bazookas and some hand grenades? Are you going to take on the Nazi army and win?" They would have said, "Across the English Channel there's a huge invasion force. It's gathering even as we speak. We don't know when the day will come, but one day they will sweep across the Channel and join us and carry us to victory." Until then, the Church militant must struggle against poverty, struggle against racism, struggle against sexism, and struggle against homophobia. But here's the good news: Even as we carry on these struggles, we are told that there is a huge invasion force being gathered beyond the sky. We do not know when it is going to happen, but one day the Church triumphant will join us and carry us to victory. *Maranatha*!

I hope that you can see that there is a social imperative in all of this. We are to work to create the kingdom of God, even now. I am not one of those old-time liberals who said that we could create it without the Second Coming. But I do believe Philippians 1:9 that says, "The one who began a good work among you will bring it to completion by the day of Jesus Christ." The creating of the kingdom has already begun, and the kingdom of God is among us. There are people who are working for the kingdom. What is initiated in us, says the Scripture, will be completed on the day of his coming! So as we labor, says Paul, we do not labor as those without hope. We labor as those with assurance that V-Day is coming.

Two illustrations, and then I'll wrap it up. I got a phone call some twenty-five years ago from a friend. He had this idea about starting an organization that would build houses for poor people, just like it says Isaiah 65. He believed that this could be done by using volunteers—people who did not know anything much about building houses. He proposed that these houses be sold to poor families who would not have to put up any payment. These families would be given long-term mortgages with no interest. He said that he would recruit these volunteer builders from churches. I said, "Yes, I'll be on your board," but I didn't expect much to happen. I like Millard Fuller. He's a nice guy. I thought it was a good thing, but never expected that twenty-five years later Habitat for Humanity would complete a hundred thousand houses, and that the next hundred thousand will be completed in just five more years. I see in this the kingdom of God breaking loose in history.

Another friend of mine, Al Whitaker, who was once the president of Mennen Corporation, created an organization called Opportunity International. This faith-based organization starts small businesses and cottage industries in Third World countries. I remember one of the first ones we started was in a slum in Santo Domingo. We got eighteen women together and started a micro-business producing sandals out of old discarded automobile tires. It worked brilliantly. It created employment that enabled people to escape from poverty with their dignity intact. We gave children fifty cents every time they brought us a discarded, worn-out automobile tire. (It wasn't long before we had every worn-out automobile tire in Santo Domingo. And then we started getting a lot of new tires!)

In just ten years, Opportunity International, has created over three and a half million jobs in Third World countries. Three and a half million jobs! And if you figure that there are six people to a family in a Third World country, you've got to take two and a half million and multiply it by six to get an idea of how many people have been delivered from poverty—not for a day, not for a week, not for a year, but for the rest of their lives. I say, "Mine eyes have seen the glory of the coming of the Lord . . . I have seen him in the watch-fires of a hundred circling camps . . . his truth—his kingdom—is marching on."[17]

Of course, I would be amiss if I did not end with the one story that's given me some mileage. It's hard to build your life on one story, but I have. I belong to a Black church in West Philadelphia. In my church, they let you *know* how you're doing when you preach, because the deacons sit right up front, and when you say something good, they yell, "Preach! Preach!" I have said many good things to you today, and none of you has cheered me on. You would all do much better if you had ten or eleven people yelling, "Preach!" every time you said something good.

The women in my church, when you say something wonderful, wave one hand in the air and they go, "Well!" That's it. "Well!" You say, "That's it?" It makes your hormones bubble, man, when you get fifty or sixty women going, "Well!" And the men in my church, when you are pumping on all cylinders, they actually stand up and point at you and yell, "Keep goin'! Keep goin' man, keep goin'! Keep goin'!" I don't get that at the C. S. Lewis Summer Institute. I don't hear anybody yelling, "Keep goin'!" I hear, "Stop! Stop!"

Once a year in my church we have a preach-off. You people don't even know what that is. In a preach-off you get about seven or eight preachers to preach back to back to see who's best. You never say that. You say, "It's for the glory of God." But, we know. And it was my turn to speak. I don't want to brag; I really don't want to brag, but I was good. I knew I was good because the deacons were yelling, "Preach!" and the women were going, "Well!" And people were yelling, "Keep goin'! Keep goin'!" And I feed on that stuff. The more they did it, the better I got; the better I got, the more they did it. They kept on doing it—I got so good, I wanted to take notes on me! I came to the end and that place exploded. There was cheering and shouting and clapping and—I mean it, it was wonderful. I sat down; my pastor hit my knee and said, "You did all right. You did all right." I said, "You're next, pastor. Are you gonna be able to top that?" He said, "Son, son, you sit back, because the old man is going to do you in."

Now, I didn't want to say anything, but there was no way anybody was going to do me in that day. But that sucker got up and for the next hour and a half he did me in with just one lousy line, over and over again: "It's Friday, but Sunday's coming." Now that doesn't sound like much, but you weren't there. He started nice and soft with, "It was Friday—it was Friday. And he's dead on the cross. But that's Friday. Sunday's comin'!" Somebody yelled, 'Keep goin', keep goin'!' That's all he needed, man. He took off. "Friday—Friday, people are saying 'As things have been, so they shall be.' You can't change history. You can't make it go in the right direction. But they don't know. It's only Friday. Sunday's comin'!"

"Friday. On Friday, darkness prevails. Evil is everywhere. There will be poverty, there will be wars, and they'll say there's nothing you can do about it, but they don't know. It's only Friday. Friday! Sunday's comin'!"

I thought I'd get something from this crowd by now. I thought maybe we could de-honkitize this C. S. Lewis crowd. I'll give you one more chance. I'll give you one more chance. It's Friday—it's D-Day, and we're living towards the future. We see the evil, but we believe that God's people are not going to limp out of history a battered, beaten church. We're going to march out triumphant because Sunday's comin'! And this is the good news. This is the good news.

He ended that sermon. I was exhausted. I was just exhausted, and he ended that sermon by yelling at the top of his lungs, as only a black preacher can, "Friday!" And with one voice, the congregation yelled back, "Sunday's comin'!"

In this view of time we can get lost in the darkness of the hour and not realize that history is going someplace. The universe has a beginning; it is moving toward a purposeful conclusion. He's coming back to establish the kingdom here on earth as it is in heaven. We have the good news in this: (Are you ready?) The good news is, it's Friday, but Sunday's comin'!

Questions and Answers

Q: Dr. Campolo, you've talked about the kingdom of earth coming, and I'm trying to get that in my brain in terms of the teaching in Peter and Revelation of the end of the world, of the destruction, of the new heaven and the new earth. How does that work out?

Campolo: I don't know. I've never been able to figure out all that eschatology stuff. It's all there, and it's yet to be reconciled, but I have learned to bracket things and say, you know, somewhere down the line I'll be able to figure out how to put all this together. I can't put a lot of stuff together right now. These systematic theologians that have a neat place for everything, I admire them. I don't make fun of them. I just admire them. I just can't put it all together yet. I see through a glass darkly. I only know in part. That's a good question, though. I would say that Peter is talking about the destruction of evil. I don't think that the God who is among us in love is not going to come back in power. Not only in Peter's writings, but also in Matthew we read that. There will come that day when all evil shall be put down, and those who embrace evil, and who have an allegiance to evil, will be destroyed. But I believe that when the fire comes, the hay, wood, and stubble of this world will be burned away; but that which is gold, silver, and precious stones will endure. When the judgment comes, I look for the world to be purified, and the destruction of all that is evil.

Q: Given the limitation of evil being in time and all that we've been discussing, and given the fact that hell itself is evil, of course, would you make a few comments if you can, about the prospect of eternity? After all, it is what Jesus came to save us from, and he knew it was a very serious thing.

Campolo: I haven't thought of that question at all, but I think it raises a whole host of intriguing implications. First of all, I'm not sure that evil is limited to time. I hate to contradict the speaker before me, but before there was ever a human race, before there was ever an Adam, before there was a fall by Adam, there was evil. Satan fell before creation, or else Satan would have never been waiting to seduce Adam. I believe there is a Satan; and I believe that Satan had a preexistence before the cosmos. If Satan had a preexistence, and if Satan was before the universe was, which the Scriptures seem to suggest, then I'm not sure that evil is limited to time. But I would hope that you're right, and that the speaker was right, that there comes that moment when evil shall be bound. I think that's what the Scripture says. Bound. Limited. Contained, so that it can no longer raise its havoc. I'm not sure that the Scripture says that Satan and the demonic, the hellish dimensions of the world, are abolished as much as they are bound and limited and they cannot do what they have been doing. There are many here who would argue otherwise, who are solid evangelicals, who say there is no hell. And there are others still who would say there is a hell, but the God that we believe in is the Hound of heaven, and descends into hell, as the Apostles' Creed suggests. It does say in 1 Peter (3:19) that Jesus goes and preaches to those in the prison house of death. The psalmist writes, "And though you descend into hell, he is there," and there are those who believe God goes on to doing saving work even there.

"I do not claim," Reinhold Niebuhr once said, "to either know the furniture of heaven or the temperature of hell, but I believe in both." I'm not sure how to handle these last two questions. I wish you would ask me something I know something about.

Q: In 1995 I was in Central Africa and I had the most amazing morning, speaking to the leaders of all the refugee camps with the Rwandan refugees. Now, 80 percent of Rwandans are Christian. The question we got to after three hours was, how was it that we who call ourselves Christian ended up killing one another? I would love to know what you would have said.

Campolo: For the same reason that we have the mess that we have in the United States, and that is that too often evangelicalism only deals with personal salvation. It deals only in terms of getting people right with God. The truth is, you're not right with God unless you're right with your neighbor. When Jesus gave the great commandment he said, "You shall love the Lord your God with all your heart, and with all of your soul, and with all of your mind" (Matt 22:37). And then he says, "And a second is like it." That is, it's the same thing: "You shall love your neighbor as yourself" (Matt 22:39). The truth is, that it is in loving one's neighbor that one loves Jesus, because the Jesus who is eternal—the cosmic Jesus—chooses to come to us through the neighbor, especially that neighbor who is hungry, especially that neighbor who is being oppressed, especially that Third World neighbor. And he says, "Come, love Me here."

On judgment day we will not be asked our views of heaven or hell. It won't be a theological test. I wish it would be, because I'm orthodox. I wish it were a theological test. I wish it was, "Before we let you in Campolo, a few questions: virgin birth: strongly agree, agree, disagree, strongly disagree." If that's what it's about, I'm in. I'm in! But, that's not what it's about. Here's what it's about: I was hungry. Did you feed me? I was naked. Did you clothe me? I was sick. Did you care for me? I was a stranger. Did you take me in? For what you did to the least of these, my brothers and sisters, you did to me (Matt 25:34-46). For the eternal Christ chooses to come sacramentally to the poor and the oppressed and bids us come love him there. And if we cannot love him there, we do not love him: For if any man says that he loves God, and does not love his neighbor, he is a liar and the truth is not in him (1 John 1:5-10).

So I think Rwandan Christians mistreated each other for the same reason that American Christians have often expressed a disconnect between being saved and being socially compassionate. It is because we have so emphasized individualistic salvation that we do not see the God of history at work in the world transforming the world that is into the world that ought to be—and calling us to do justice in the world.

Q: In your talk, you mentioned Stephen Hawking and his current explorations of ideas about a cyclical pattern of the creation of earth and matter, and you seemed to say that if he comes up with empirical evidence, we should throw the Bible out, because he finds it actually is cyclical according to what he knows now, or what the evidence suggests. It seems like we are in a problem area. Darwin challenged a lot of ideas that the church held onto, and the church practically collapsed. Not really, but in a sense it did. I would hate to think that the church is hinging on what science does, to the degree that if science comes up with an idea that doesn't fit our pattern, we no longer can stand faithfully with our beliefs.

Campolo: You have hit on a very important theme. Whenever our theology is wrapped up with a specific cosmology, we are always in danger, because the cosmologies change even though we think they will never change. What seems to be unquestioned today is doubted and actually discarded tomorrow. So, if our theologies are contingent upon any given scientific cosmology, we are in dangerous straits. And yet, that is the risk that we have to take. If we are to communicate the gospel within a given existential situation, we have to, in fact, deal with the issues that emerge in that situation. But even as we do so, as your question suggests, we must do so with great humility, recognizing that this isn't the final word. If Hawking comes out right, then we'll have to go back to the drawing board and come up with a new way of thinking. We must not assume that our theologies are so set in stone that they cannot be challenged, "because the times, they are a-changing," and any theology that is tied to a specific cosmology is in danger of being doomed, as your question suggested.

As a sociologist, however, I would argue that one has to, in fact, relate the gospel in the existential situation. In short, we have to make the gospel make sense in the terms in which people are speaking. Paul certainly did that when he preached the gospel on Mars Hill and referred to Epicureans and Stoics (Acts 17). He was attaching the message to an existentialist philosophy that was there and then. Those philosophies are now out of vogue, but Paul used them to make his arguments. What we must be ready to do is to be willing to recast our theology in light of new cosmologies.

Thank you very much. You've been very attentive and very good to me.

Conclusion

Harry Lee Poe

When the scientific, philosophical, and theological worlds percolate with alternative theories about the nature of time, God's relationship to time, and God's knowledge of the future, the most we can hope to do is make a few preliminary, tentative conclusions. These reflections do not represent a consensus of the contributors to this volume. The contributors remain as innocent as lambs. At the end of the day, theologians, pastors, and laity who have no particular expertise in the technical areas that have informed this collection of essays will decide what they believe about the issues raised in this book.

The issues for the average person sometimes differ from the issues of the specialist. Sometimes, however, the specialist works on a problem that has direct bearing on the everyday concerns of life, but the average person, the pastor, and even the professional theologian may fail to recognize the connection. When I finished my theological studies, I first served as pastor of a church. I recall the first time a grieving widow asked me *where* her husband was *now*. I have been asked the question many times since, and I have realized that the question concerns both time and space. The other question that the widow asked that I have heard many times since concerns whether or not her husband *knows* she loved him. For a Christian, these questions lie at the intersection of biblical faith and modern scientific and philosophical understandings of the world.

Each religion views time from a unique perspective and brings different questions to science about time. Families of religions may ask similar questions, but they also have distinct concerns as well. The three monotheistic religions have a common notion of the flow of time rooted in the understanding of a Creator who guides creation toward a final goal. In the same way, the reincarnational religions that emerged from India have a different perspective on time that relates to their understandings of the progress of reincarnation. The two families of religions (monotheism and reincarnationalism) have different perspectives on the nature of the physical world that science studies.

The meaning of time first entered my mind as a serious theological issue during my seminary days when one of our New Testament professors remarked in chapel that God does not know the future. This statement startled me, because I had always thought that God knew everything. During these early days of my Master of Divinity program I began to be aware of the diversity of opinion within the Christian scholarly world about matters theological.

The meaning of time entered my mind a second time as a serious theological issue when I first read Genesis in Hebrew and realized that the grammar and verb forms did not support my hitherto literalistic understanding of creation as the activities of God over the course of a work week. This discovery helped me understand the difference between revelation from God and the *interpretation* of revelation from God.

The meaning of time entered my mind a third time as a serious theological issue when I chanced to pick up Nigel Calder's little book, *Einstein's Universe*, on the discount table of a local bookstore during my doctoral studies. I had developed an interest in science and religion through a course I had taken with Eric Rust, one of the few Christian philosophers working on science and religion in the 1950s and 1960s. What struck me about Einstein's understanding of time and space was the implication for the nonphysical. If time is an aspect of the physical world, then it would seem that it is not an aspect of the nonphysical world. The more I read, however, the more I realized that physicists and theologians either did not share my conclusion, or they were unaware of the issue.

The New Testament professor who first suggested to me that God does not know the future lived in a Newtonian universe. Time for him was an absolute metaphysical reality. While he was very modern in most of his thinking, he knew little of science. He was one of a vast multitude of American Christian theologians influenced by Karl Barth's brand of neo-orthodoxy that had no interest in science. They represented a modern strand of Reformed thought while accepting little of the content of Reformed thought. Calvin had dismissed the possibility that general revelation is sufficient by itself to lead to salvation, and the great Reformed tradition since then has seen very little value in science as a handmaiden to theology. General revelation was taken up later by the Deists, however, who had little use for specific revelation. Through the influence of the Deists, general revelation became the basis for natural theology that dominated English Protestant thinking about science and religion from Newton to Darwin. Barth and his followers had no interest in science because they tended to think of it solely in terms of the brand of natural theology that prevailed in England in the late eighteenth and early nineteenth centuries. If it did not provide a path to salvation, then it had no value for informing theological thought.

At the end of the twentieth century, the question of what God knows became an issue among American Evangelical theologians and philosophers, just as it had been among mainline Christians earlier in the twentieth century. John Sanders, Clark Pinnock, Greg Boyd, and several other evangelical theologians have proposed that God does not know the future—that God is "open" to the future—in contrast to the traditional Christian understanding. For the open theists the issue at stake is the problem of pain.[1] They reason that God would surely intervene if he knew something bad was about to happen. For the traditionalists, the issue at stake is the very nature of Scripture and the significance of prophecy. If God does not know the future, then there can be no prophecy that is fulfilled. Both sides are motivated by extremely important issues which neither side seems willing to address. Those concerned with theodicy appeal to emotion and those concerned with revelation appeal to tradition. Oddly enough, within the evangelical framework, neither emotion nor tradition provides an adequate authority. At the same time, however, neither side in the debate seems interested in the question of God's relationship to time.[2]

If Einstein was right about time and it is a fourth dimension of physical reality, then time is limited to the physical realm. For materialistic naturalists, this state of affairs raises no problem, because for them, nothing exists that is not physical. For a theist, however, time as a function of space suggests that God exists outside of *time* just as God exists outside of *space*. Not all religions ask the same questions of science, however, since they have different understandings of the world. Relativity does not raise the same questions for Hinduism and Buddhism because of their concepts of physical reality and its relationship to the divine.

The theist also insists upon God as a personal being. Paul Davies has argued that God cannot be timeless and personal, since thinking, conversing, feeling, planning and other activities associated with persons are all temporal activities. In his argument, however, Davies commits the anthropomorphic error of confusing being a person with being human. In so doing, he projects his own human experience on the Godhead.

Hugh Ross offers another proposal with his multidimensional view of God who is not timeless, but who exists in a dimension of hyper-time. As soon as we speak of dimensions, however, it seems that we have definitely entered the physical realm. Ross's analysis makes clear that whether theists conceive of God within or without time, they must still account for how God relates to the physical world.

Process Theology works from the other end of the issue as it brings God not only into time, but also into space. While Process Theology has as much diversity as Protestantism, an emerging trajectory of contemporary Process thought appears to be moving toward an identification of God with Nature or the Process. We see this trend especially as other voices concerned with ecology and feminist issues merge with the old Nature Goddess spirituality. This direction would represent a polar opposite to an Augustinian interpretation of the space-time problem in light of Einstein. While an Augustinian position may be stated that since God is spirit, God is not within space or time, a Process position might be stated that since God is within time, God is also within space. The Process Theologians are the heirs of the Deists in the sense that specific revelation and Christian doctrine do not play a definitive role in how they approach their subject. From a methodological perspective, the Reformed tradition and the Process tradition are virtually identical. The Reformed approach relies upon the tradition without engaging contemporary insight. The Process approach relies upon contemporary insight without engaging tradition. The Reformed approach forgets that tradition began as contemporary insight long ago. The Process approach forgets that contemporary insights of yore were frequently wrong.

When we consider the Big Bang implications of general relativity, we have a beginning to time and space. What happened before the Big Bang? If Einstein is right about time and space, and if the Big Bang is the beginning of time and space, then it is inappropriate to speak of anything *before* the Big Bang. There would be no *before* the beginning of time. Stephen Hawking's efforts to discredit the Big Bang theory relate to his argument that if the universe did not have a beginning, then there is no need for a creator. His approach depends upon his concept of "imaginary time." Of course, the problem of the existence of the universe has been around since long before the Big Bang theory. Even if the universe is eternally subsistent, "just because" has never been a satisfying explanation for children, much less for scientists, philosophers, and theologians.

As it is, not everyone agrees with Einstein about time and space. John Polkinghorne argues that time must also have a *metaphysical* counterpart apart from space. William Lane Craig has argued in *Time and Eternity* for Hendrick A. Lorentz's interpretation of the special theory of relativity over against Einstein. By Lorentz's interpretation, absolute time and space actually exist, but they cannot be measured because measuring instruments contract and/or slow down relative to the motion of what is being measured. Craig favors the view that God is timeless without creation but temporal since creation. Craig affirms God's foreknowledge by the theory of middle knowledge whereby God's omniscience extends to all future contingents.

Even with Einstein's understanding of time and space as a single physical reality, time and space have been interpreted differently. The most common interpretation involves "flowing time" which is a traditional Western unidirectional understanding of the "arrow" of time. It involves a beginning and an end. In Hebrew thought it involves a goal or end point. Russell Stannard defends a second view of time and space known as the "block universe" that conceives of time and space as a complete reality from beginning to end in which all points of time exist. In flowing time, the present moment has the privileged position as the standard point of reference, but in the block universe no point in time has privilege. They all exist. In flowing time, the past is gone and the future has not yet arrived. In the block universe, all points exist and all points are the present with respect to themselves.

The block universe is not appreciably different from Augustine's understanding of how God relates to time and space. Augustine conceived of time as a creation like space. God exists apart from time just as God exists apart from space. From the divine vantage point, all of time from beginning to end lies unfolded like a scroll. All points are simultaneous for God, and yet no points are simultaneous with God. One need not be a Process Theologian or a panentheist to find this view threatening to a doctrine of human freedom. Old-fashioned Wesleyans may also take exception to this view.

Both flowing time and block time present a problem for understanding the present moment. Timothy George, Russell Stannard, John Polkinghorne, William Lane Craig, Robert Russell, and Tony Campolo have explored the problem of the reality of the present moment. George points out that Augustine struggled with the problem of measuring "now." Stannard has argued that no moment of time is privileged with respect to another in Einstein's universe. Polkinghorne, on the other hand, while exploring options for understanding time, argues for a "natural global cosmic 'now'" that represents the "preferred point of view from which to think about cosmology." Craig points out the problem with speaking of a "now" in the experience of God, if God stands outside of time. In his discussion of eschatology, Russell argues for a "copresence" of all events and suggests that all events have a "thickness" to them instead of the "point-like" conception that people tend to have of now. Finally, Campolo focuses on the existential experience of the present and a reading of Einstein that suggests every moment of time is simultaneous to God who stands outside of time and experiences neither past nor future.

While most physicists see flowing time and block time as the two options for understanding time in relation to modern physics, not all physicists agree. Julian Barbour represents a minority view (not explored elsewhere in this collection) that time does not exist. Whereas most physicists struggle with the concept of the present moment, Barbour insists that only the "instants of time" exist. The flow of time is an illusion created in the human brain. The instants of time, or the all the possible "nows," exist as "time capsules" of static records. Change does not occur. The brain organizes the most likely possible "nows" in such a way that the illusion of change and the flow of time present themselves to the consciousness.[3]

The problem of identifying or measuring the present moment of time has a striking parallel with the problem of finding the location of an electron when measuring its velocity. The wave/particle duality of quantum mechanics has its corollary with the flow/moment of time. If Einstein is correct about time as an extended dimension of space, it should not be surprising to see time behave similarly to matter. Time extends or "flows" like a wave. We cannot identify or measure the present moment (the "now") when measuring time any more than we can identify the location of an electron while measuring its velocity. The problem of locating and measuring "now" is that it corresponds to the attempt to know the electron as a particle when measuring it as a wave. We may reasonably ask if it is possible for the human observer, or any physical observer, to ever locate the "now" of time. As part of the space-time continuum, any physical observer exists in the flow of time just as the physical observer exists physically. We can measure both the point and the wave of the electron because we stand outside the atoms we measure. We do not stand outside time. We are part of it.

A reciprocal relationship exists between the movement of time and space in the measurement of one another. We measure the passage of time by the motion of physical matter: the movement of the earth, moon and sun, the passing shadow of the sun, the movement of the hands on a clock face, the vibrations of an atom. Likewise, we measure the movement of space by time: miles per hour, knots per twenty-eight seconds, miles that light can travel in a year.

Unlike space, time is invisible. We do not see it, but we infer it. We do not confuse the measurement of space with space itself, but we may confuse the measurement of time with time itself. The present moment in time is "now" while the present place in space is "here." We have the same problem with identifying "here" as we do with "now." "Here" may represent a point on a physical line just as "now" represents a point on a time line. The problem with the present moment, however, lies in its resemblance to Leibnitz's infinitesimal, that disquieting feature of his calculus that always allows for yet a smaller division of a number. The present moment "now" can always be divided into a smaller moment. Because we can see "there" as well as "here," the problem with having precise knowledge of "here" is not so acute for us. With time, however, we can see neither "now" nor "then." To the scientific problem of certainty about "now," we must add the existential problem. Perhaps the existential condition of the observer is the reason that certainty over "now" creates such a problem long after physicists have come to grips with the Heisenberg Uncertainty Principle for quantum mechanics.

The fact that we measure time (*chronos*) suggests that something must exist that we measure, but it may all be a mental construct. The bases for measurement are certainly mental constructs in some cases: seconds, minutes, hours. Yet, in other cases, the bases have to do with the observation of motion or change: the day (the rotation of the earth) and the year (the orbit of the earth around the sun). The fact that we do measure, however, suggests the existence of something that allows the observation of motion and change. We measure space at rest through mental constructs: centimeter, meter, kilometer. Yet, in other cases, we measure the physical world by comparison with other matter in the physical world: the length of a king's foot or the width of pharaoh's palm. Still, we do not confuse the basis for measurement with the physical world we measure. We do not confuse fathoms and leagues with the sea itself. We do not confuse inches, feet, and rods with the earth itself.

The Trinity

While the issues for theists differ from those of Buddhists and Hindus, the issues for Trinitarians may provide a solution for how God relates to time and space. William Lane Craig helps us recognize that the creation of time and space introduces a radical new experience for God in terms of relating to something other than God. This experience of relating to time and space rivals the incarnation in terms of relating to something radically different from God. Philosophical discussions about God and time, however, rarely involve a discussion of the Trinitarian God who exists as Father, Son, and Holy Spirit.

Christians affirm a single Creator God who exists in three persons. At the same time, they affirm both the complete transcendence of God who is totally other than the physical world and the true incarnation of God for thirty-three years as fully God and fully human. The incarnation suggests that the persons of the Godhead relate to time and space in different ways. In time and space, the Son of God experiences the full affect of physical existence unless ameliorated by God the Father, including limited knowledge of the future. God the Holy Spirit continues the divine involvement in creation and mediates God's will over creation. This mediating role of the *paraklete* involves the intercession between God and the physical order.

The problem of the Trinity involves, among other things, the problem of divine knowledge. Just as God's knowledge concerns time as an aspect of physical reality, the Trinity relates to space as a physical reality. Christians understand God to consist of three persons in a non-physical, non-spatial, non-temporal state of being. The problem of the Trinity exists from the perspective of a spatial and temporal world in which we would tend to conceive of three persons as discreet beings.

The attributes of God apply to all three persons of the Godhead—attributes such as love, holiness, and goodness. Distinguished from the attributes, however, are the properties of the persons of the Godhead. Omniscience, omnipotence, and omnipresence are properties ascribed to God, but they do not necessarily apply to all three persons of the Godhead *in time*. For instance, the Son did not possess these properties during the incarnation, but that thirty-three year temporal experience does not preclude these properties before or after the incarnation. One might argue that the Holy Spirit possesses the property of omnipresence in relating to the entire physical realm. On the other hand, one might argue that the Father does not exist everywhere, for "everywhere" would suggest physical location. Instead, the Father might be said to exist "no *where*." As the exalted Lord who governs the universe, the Son functions in a relationship to the physical world distinct from that of the Father who does not step into this relationship until the new creation. The Son mediates between the world of humans and the transcendent Father, while the Holy Spirit mediates the world to the Son.

The Father

In the Hebrew and Christian Scriptures, God does not appear to anyone except "the innocents" in Genesis, which suggests an experience out of time.

Otherwise, Yahweh reveals himself and acts through angelic messengers and through the Spirit of the Lord, and the Father acts and speaks through angels and through the Holy Spirit. The Godhead, Yahweh, the Father, the Ancient of Days, remains constantly aware yet forever removed from the world of time and space. The Father is completely transcendent, eternal in being, complete in perfection, absolute in holiness and all other attributes. As such, the Father does not relate directly to the immanent, the temporal, the incomplete, or the tentative. Would this approach suggest an atemporal modalism? The dynamic relationship of Father, Son, and Spirit across the divide—of heaven and earth, eternal and temporal, spiritual and physical, transcendent and immanent—would mitigate against any form of modalism.

The Son

The incarnation provides a case study for how God's knowledge does not destroy human free will. In the complex dynamics of the cross, the Father knew and experienced the decisions of the Son who could not see beyond the moment in time of the flesh, except when the Father parted the veil of eternity and made things known to him. The incarnation meant captivity to the space-time continuum or no real incarnation would have taken place. Jesus could be killed, and he could not know the future, but the Father raised him from the dead and made the future known to him.

The Son not only entered into space in the incarnation; he entered into space-time. Even if the Father experiences time in some other way, the Father would not experience it as space-time. To do so, the Father would also have to take on physical form and experience incarnation into the space-time continuum.

The Son declared that no one knows the day or the hour of the events that lead up to the end of time itself (Matt 24:36), except the Father. On the other hand, the Son did know what was in the human heart (Mark 2:8; Luke 5:22), which involved knowledge of the present moment contemporary with the incarnation. The incarnation involved spiritual knowledge which Jesus perceived because the incarnation did not impose limits upon the divine nature: he enjoyed full holiness, love, joy, peace, and the plethora of other attributes of the divine nature not limited by time and space.

The Holy Spirit

The Holy Spirit of God proceeds from, but is not separate from, God the Father, the Ancient of Days. The Holy Spirit extends into time and space and exercises the power of God in time and space. In the first mention of the Spirit in Genesis, the Spirit of God moves upon chaotic creation before it is ordered. The Holy Spirit is not a servant or lesser being, like an angel. Neither is the Holy Spirit another spiritual being, like the demiurge of Platonism. The Holy Spirit is God involved in the space-time continuum, while the Father is God who remains eternal. While the Holy Spirit exercises the power of God and shares all attributes of God, the Holy Spirit never acts or speaks in Hebrew or Christian Scriptures apart from the Father or the Son. Whereas the incarnation of the Son has a particular character, coming to a particular place and point in space-time, the Holy Spirit has an extended character, like a wave.

The relationship of the Father, Son, and Holy Spirit suggests a solution to how God relates to the physical world of time and space. The Holy Spirit interacts with the world of phenomenal experience on a temporal basis and mediates the physical world to the Son who guides the cosmos. The Son, in turn, interacts with humans through the Holy Spirit and mediates with the Father who exists beyond time and space. As the exalted Christ, the Son reigns over the universe and the unfolding of history. The Son guides the process of redemption to its conclusion: "When all things are subjected to him, then the Son himself will also be subjected to the one who put all things in subjection under him, so that God may be all in all" (1 Cor 15:28). At the new creation, the nature of reality changes, but God does not change. In the dazzling imagery of Revelation, the Father and the Son sit on a single throne and the Holy Spirit (the river of the water of life) proceeds from them (Rev 22:1). The apparent contradictions of the Triune God disappear as the nature of space-time changes in the new creation.

THE ESCHATON

Robert Russell's speculations about eschatology and its implications for both science and Christian theology suggest the possibility of "reverse causality" from future to present, rather than the unidirectional causality of past to present with which we have grown accustomed. Such an experience as "reverse causality" might also be construed as coming from outside time rather than from the future. Russell rightly associates the event of the resurrection of Christ with eschatology and the "new creation." The resurrection involves continuity with the present experience of the world, yet it involves a radical transformation of the old into something amazingly new. Do we have grounds to expect a "new physics" of the resurrection?

The experience of the universe is a story of changing physics. The gradual emergence of the four forces (did they emerge, or did they separate?) as energy begat matter suggests that the laws of physics have changed and may change again. It does no good to say that the four forces (weak nuclear force, strong nuclear force, electromagnetism, and gravity) emerged quickly in scant "moments" of time. Whether the four fundamental forces began as a unified field or gradually emerged does not matter for purposes of illustrating that the physics of the universe has changed. From our perspective, the individuality of the four forces took place so long ago that it is rather easy psychologically to dismiss their significance for a universe whose physics may change again.

By recent reckoning, gravity emerged as an individual force at 10^{-43} seconds after the Big Bang. For those not used to dealing with large and small numbers, this number represents a fraction of the first second of time with a decimal followed by 42 zeros and a one:

.000000000000000000000000000000000000000001

At 10^{-35} seconds the strong nuclear force emerged and the great inflation occurred, dramatically increasing the size of the universe from the size of an atom to the size of a cherry pit. The number of zeroes is less, but still it is a tiny fraction of our time:

.00000000000000000000000000000000001

From our perspective, the passage of time between the two events is as insignificant as the miniscule size of the new universe. If we go back in time to the emergence of gravity, however, then the universe would have been around since the beginning of time! It was not a short amount of time then; it was a huge amount of time. Likewise, the space of the universe, though only the size of a cherry pit, had suddenly enlarged enormously. We speak of the passage of time in terms of tiny fractions of a second, but there were no seconds, days, or years then, only the passage of time and the expansion of space up to that point. The universe had always been the same up to that point, then the physics changed. From our perspective, it would be the equivalent of the universe behaving the same way from the stabilization of the four fundamental forces at 100 billionth of the first second until now. At least, it would seem logical that the relative nature of time would work that way at the beginning, except that before the four fundamental forces operated as discrete forces, the physics of relativity did not exist as it does now. It is neither logically nor scientifically unreasonable to expect that our universe will have other physics in its future. What sort of situation might cause such a change? It is hard to say since we do not know what sort of situation caused the last great change.

DEATH AND SALVATION

When Christ died and entered Paradise, the realm of the righteous dead, he preached good news to the spirits in prison (1 Pet 3:18-19; 4:6). Given what Einstein taught about the relationship between time and space, and given that death involves radical cessation of physical, embodied existence, the state of death involves no necessary temporal relationship to the physical world. If these rather large assumptions are true, then Christ may have encountered all the dead of all time in his death. Such a view, however would seem to require some form of a block understanding of time, but the simultaneity of the experience of the dead lies outside of time.

Christians believe that salvation involves incorporation into Christ through faith. Those who give themselves to him and are incorporated into him survive death because they are in him, the one who conquered death through resurrection. One of the perpetual questions about the justice of the Christian understanding of salvation concerns those who have never heard of Christ. The experience of time for the dead may shed light on this issue. Christians allow that Abraham and Sarah, Isaac and Rebecca, and all the people of faith before the incarnation of Christ share in salvation, because when Christ died, he descended into the realm of death and set the captives free (Eph 4:8-10).

The Apostle Paul states that Abraham believed the gospel beforehand, but all he knew to believe was that through his descendant, all the world would be blessed (Gal 3:6-9). Peter tells us that the prophets spoke of the sufferings of Christ, but that not even the angels understood the message beforehand (1 Peter 1:10-12). If death is a timeless state, then those who died before the crucifixion of Christ as well as those who die after the crucifixion have a simultaneous experience of death. Those who have never heard the gospel of Christ may be included in the number of righteous dead who believed what little they knew of God and it was counted as righteousness.

Thus, the resurrection is an eschatological event. Christ raises the dead at the end of time with him, but he appears in the middle of time for the benefit of his disciples. Because the dead do not exist in a physical state or inhabit a physical location, no passage of time (at least, time as we know it) occurs for the dead. This understanding of the timelessness of death may account for how the apostles spoke of the dead in Christ being asleep and yet being immediately with Christ. The body may wait in the dust for thousands of years to experience resurrection, while the human spirit in a timeless state goes immediately from death to life.

In a book composed of lectures that came from the 2002 C. S. Lewis Summer Institute, it may be appropriate to comment on C. S. Lewis's fascination with space-time and its implications for theology. In his stories of Narnia, Lewis tells a story in which different worlds experience time as a relative thing. In *The Magician's Nephew*, children pass from one world to another through pools of water, but these are devices of children's fantasy that serve to explore the possibility of worm holes that connect vast outreaches of the universe or even parallel universes. Lewis saw no problem with multiple universes. The more the merrier. In *The Last Battle*, Lewis considers the problem of time and space for death. In the story, people die at different times and places in different worlds, but they all appear before the great lion Aslan "at the same time."

CONCLUSION

We may dismiss the scientific and theological views of C. S. Lewis since he was not a scientist or a theologian, but if we do so, we miss the point. The advances in scientific understanding have not come through reliance upon the conventional way of understanding things. Every advance has been a break with the old understanding. The same could be said of theology, but both disciplines rely upon an unshakeable foundation that allows for the examination and rejection of old theories. Regardless of which cosmology or theory of quantum mechanics will prevail, science clings to the belief that the universe is knowable, that it has an order to it that allows for prediction regardless of how chaotic its systems may be, and that those who seek shall find. Theology clings to the belief that God exists, that God created the universe, and that those who seek shall find.

It is possible to become too isolated within our own domain, however, to gain the spark of insight necessary to begin a new line of inquiry. Before we dismiss the insights of a professor of medieval and renaissance literature, a man who wrote children's stories and science fiction, let us recall how dismissing such people in the past has retarded the discovery of new knowledge.

The creator of the science fiction story first proposed the relativity of time some seventy years before Einstein ("Three Sundays in a Week"). He proposed a Big Bang theory some eighty years before Lemaître (*Eureka*). He also proposed downward or mutual causality a hundred years before Austin Farrar. People called Edgar Allan Poe a madman for rejecting absolute time, a static universe, and a deterministic understanding of cause and effect. The germ for new knowledge does not always lie in the same played-out mine from which all the gold has been extracted. The constant interaction between fields may best stimulate us to ask questions in a different way.

Notes

Chapter One

1. Jameson proposed his theory in the preface to his English translation of Cuvier's *The Theory of the Earth*. See M. Cuvier, *Essay on the Theory of the Earth with Mineralogical Notes and an Account of Cuvier's Geological Discoveries by Professor Jameson* (Edinburgh: William Blackwood, 1817).

2. Buckland's Gap Theory appears in his *Relics of the Flood*. See William Buckland, *Reliquiae Diluvianae; or, Observations on the Organic Remains contained in Caves, Fissures, and Diluvial Gravel, and on Other Geological Phenomena, Attesting the Action of an Universal Deluge* (London: John Murray, 1824).

3. Westermann traces the modern discussion to Rabbi Solomon ben Isaac (d. 1105), who argued for "When God Began . . . the earth was . . . empty. . . ." For a discussion of the options put forth, see Claus Westermann, *Genesis 1–11*, trans. John J. Scullion (Minneapolis: Augsburg Publishing House, 1984), 94–97.

4. Spurrell explains, "If we remember that the tenses in Hebrew do not indicate the *date*, but the *state* of an action, i. e. whether it be *complete* or *incomplete*, the

explanation of this peculiar Hebrew construction will not be far to seek. The impref. denotes an action as entering on completion. When we have a series of events, each single event need not necessarily be regarded as completed and independent, but each may be regarded as related to the preceding one, one event stepping into its place after the other, the date at which each successive event comes in being determined by the ו [waw], which connects the new event with a point previously marked in the narrative" (*Notes on the Hebrew Text of the Book of Genesis* [Oxford: Clarendon, 1887], 5–6).

5. The difference of meaning is a particular problem for those who defend a twenty-four-hour solar-day understanding of the passage. The absence of the definite article is virtually without precedence in the biblical text. Andrew E. Steinmann has presented a theory that the choice of "one day" in 1:5 rather than "the first day" suggests that the text is defining a day as morning and evening, or a solar day. This theory, however, fails to account properly for days two through five without the definite article. See Andrew E. Steinmann, "אחד as an Ordinal Number and the Meaning of Genesis 1:5," *Journal of the Evangelical Theological Society* 45 (2002): 577–84.

6. C. I. Scofield, "Introduction," *Scofield Reference Bible* (Oxford: Oxford University Press, 1945), iii.

7. Edgar C. Whisenant, *88 Reasons Why the Rapture Will Be in 1988* (Nashville: World Bible Society, 1988), 1.

8. Ibid., 7.

9. George Beasley-Murray has made a similar observation about the judgments of the trumpets and the bowls. See G. R. Beasley-Murray, *The Book of Revelation*, New Century Bible Commentary (Grand Rapids: Eerdmans, 1983), 238–39.

10. Ibid., 129–30. Beasley-Murray stands in a long tradition of expositors who recognize the relationship between the little apocalypse and the seven seals.

11. C. S. Lewis, "De Descriptione Tempore" (Cambridge: Cambridge University Press, 1955); reprinted in *Selected Literary Essays* (Cambridge: Cambridge University Press, 1979).

Chapter Two

1. St. Augustine, *The Confessions*, trans. Maria Boulding (Hyde Park, NY: New City Press, 1997), 11.1.1.

2. C. S. Lewis, *Mere Christianity* (San Francisco: HarperCollins, 2001), 166.

3. Ibid., 171.

4. Jaroslav Pelikan, *The Mystery of Continuity: Time and History, Memory and Eternity in the Thought of Saint Augustine* (Charlottesville: University Press of Virginia, 1986), 35.

5.   Friedrich Schiller, "Spruche des Confucius," in Jehiel K. Hoyt, *Hoyt's New Cyclopedia of Practical Quotations from the Speech and Literature of All Nations* (New York: Funk & Wagnalls, 1940), 798a.

6.   *Conf.* 11.14.17. See also editor's comment on this passage.

7.   Adrienne Rich, "The Desert as Garden of Paradise," in *The Fact of a Doorframe: Selected Poems 1950–2001* (1989; New York: W&W Norton, 2002), 29–30.

8.   Friedrich Nietzsche, *Thus Spoke Zarathustra* (New York: Penguin, 1969). See the helpful exposition of Nietzsche in Stephen Kern, *The Culture of Time and Space* (Cambridge, MA: Harvard University Press, 1983), 52–86.

9.   *Conf.* 11.15.20.

10.   Emil Brunner, "The Christian Understanding of Time," *Scottish Journal of Theology* 4 (1951): 1–12.

11.   Friedrich Hölderlin, "Hypherion's Song of Fate," in Karl Barth, *Church Dogmatics* 3/2 (Edinburgh: T&T Clark, 1960), 515.

12.   *Conf.* 11.16.21.

13.   Friedrich Schleiermacher, *On Religion: Speeches to Its Cultured Despisers* (1799). See *Cambridge Texts in the History of Philosophy*, ed. and trans. Richard Crouter (Cambridge: Cambridge University Press, 1996).

14.   *Conf.* 11.13.16. See the excellent commentary by R. J. O'Connell, *St. Augustine's Confessions: The Odyssey of Soul* (New York: Fordham University Press, 1989), 135–44.

15.   Boethius, *The Consolation of Philosophy*, 5:6.

16.   St. Thomas Aquinas, *The Summa Theologica*, Part One, q10, a.2: "*Nec solum est aeternus, sed est sua aeternitas: cum tamen nulla alia res sit sua duratio, quia non est suum esse. Deus autem est suum esse uniforme: unde, sicut est sua essentia, ita est sua aeternitas.*"

17.   *Conf.* 11.8.15.

18.   T. F. Torrance, *Space, Time and Incarnation* (New York: Oxford University Press, 1969), 55.

19.   Augustine, *On the Holy Trinity in Nicene and Post-Nicene Fathers*, ed. Philip Schaff (Peabody, MA: Hendrickson, 1887), 5.1.2.

20.   The Jewish midrash is recounted in Pinchas Lapide and Jürgen Moltmann, *Jewish Monotheism and Christian Trinitarian Doctrine* (Philadelphia: Fortress, 1981), 65.

21.   Cited in Seyyed Hossein Nasr, *Islamic Spirituality* (New York: Crossroad, 1987), 321.

22.   This theme is expressed in James Weldon Johnson, *God's Trombones* (New York: Viking, 1948).

23.   Anthony Towne, *Excerpts from the Diaries of the Late God* (New York: Harper & Row, 1968), 11.

24.   Karl Barth, *Church Dogmatics* 4/2 (Edinburgh: T&T Clark, 1960), 755.

25.   T. S. Eliot, "The Rock," in *Collected Poems, 1909–1935* (London: Faber & Faber, 1936), 173.

26.   Thomas Hardy, "The Dynasts," in *The Works of Thomas Hardy in Prose and Verse* (London: Macmillan, 1913), 2:254.

27.   Barth, *Church Dogmatics* 2/1, 616.

28.   Friedrich Nietzsche, "The Gay Science," (1882), 341 in *The Portable Nietzsche* trans. Walter Kaufman (New York: Viking, 1954), 101–2.

29.   The Venerable Bede, *A History of the English Church and People* (New York: Dorset, 1985), 186.

30.   Helen H. Lemmel, "Turn Your Eyes Upon Jesus," 1922, renewed 1950 by H. H. Lemmel. Assigned to Singspiration, Inc.

31.   See comparison of Croly and Lemmel in Mark Noll, *Scandal of the Evangelical Mind* (Grand Rapids: Eerdmans, 1994), 144–45.

32.   Dante, *The Divine Comedy* (New York: Oxford University Press, 1948) *Paradisio* 33:133–45. On the image of Christ in the Salamanca cathedral, see Geoffrey Wainwright, "Sacramental Time," *Studia Liturgica* 14 (1982): 135–46.

Chapter Four

1.   This section is based on John C. Polkinghorne, "Natural Science, Temporality, and Divine Action" in *Theology Today* 55 (1998): 329–43. See also, John Polkinghorne, *Faith, Science, and Understanding* (New Haven: Yale University Press, 2001), ch. 7.

2.   D. Bohm and B. J. Hiley, *The Undivided Universe* (London: Routledge, 1993).

3.   John Polkinghorne, *Belief in God in an Age of Science* (New Haven: Yale University Press, 1998), ch. 3; Ilya Prigogine, *The End of Certainty* (New York: The Free Press, 1996).

4.   Polkinghorne, *Belief in God.*

5.   See, Arthur Peacocke, *Creation and the World of Science* (Oxford: Oxford University Press, 1979).

6.   John Polkinghorne, *Science and Providence* (London: SPCK, 1989), ch. 7.

7.   Robert J. Russell, Nancey Murphy, and C. J. Isham, eds., *Quantum Cosmology and the Laws of Nature* (Vatican Observatory/CTNS, 1993).

8.   John Polkinghorne, *The God of Hope and the End of the World* (New Haven: Yale University Press, 2002).

Chapter Five

1.   See Eleonore Stump and Norman Kretzmann, "Prophecy, Past Truth, and Eternity," *Philosophical Perspectives* 5 (1991); cf. idem., "Eternity, Awareness, and Action," *Faith and Philosophy* 9 (1992); Paul Helm, *Eternal God* (Oxford: Clarendon, 1988); Brian Leftow, *Time and Eternity*, Cornell Studies in the Philosophy

of Religion (Ithaca: Cornell University Press, 1991); John Yates, *The Timelessness of God* (Lanham, MD: University Press of America, 1990).

2.　　　Alan G. Padgett, *God, Eternity, and the Nature of Time* (New York: St. Martin's, 1992); Richard Swinbourne, *Space and Time*, 2d ed. (New York: St. Martin's Press, 1981); Stephen T. Davis, *Logic and the Nature of God* (Grand Rapids: Eerdmans, 1983); William Lane Craig, Paul Helm, Alan Padgett, and Nicholas Wolterstorff, *God and Time: Four Views*, ed. Gregory Ganssle (Downer's Grove: InterVarsity Press, 2001).

3.　　　Laura Ingalls Wilder, *Little House in the Big Woods* (New York: Harper & Row, 1932), 237–38.

4.　　　William Lane Craig, *Time and Eternity: Exploring God's Relationship to Time* (Wheaton: Crossway Books, 2001).

Chapter Six

1.　　　Adapted from Robert John Russell, "Eschatology and Physical Cosmology: A Preliminary Reflection," in *The Far Future: Eschatology from a Cosmic Perspective*, ed. George F. R. Ellis (Philadelphia: Templeton Foundation Press, 2002), 266–315, with permission from the Templeton Foundation Press, from Robert John Russell, "Bodily Resurrection, Eschatology, and Scientific Cosmology: The Mutual Interaction of Christian Theology and Science," in *Resurrection: Theological and Scientific Assessments*, eds. Ted Peters, Robert John Russell, and Michael Welker (Grand Rapids: Eerdmans, 2002), 3–30, with permission from Eerdmans Publishing Company, and from Robert John Russell, "Sin, Salvation, and Scientific Cosmology: Is Christian Eschatology Credible Today?" in *Sin & Salvation*, eds. Duncan Reid and Mark Worthing (Australia: ATF Press, 2003), with permission from the Australian Theological Forum.

2.　　　Ted Peters, *God as Trinity: Relationality and Temporality in the Divine Life* (Louisville: Westminster/John Knox Press, 1993), 175–76.

3.　　　See n. 1 above.

4.　　　John Macquarrie, *Principles of Christian Theology*, 2d ed (New York: Scribner's Sons, 1977 [1966]), ch. 15, esp. pp. 351–62.

5.　　　For a nontechnical introduction, see James Trefil and Robert M. Hazen, *The Sciences: An Integrated Approach*, 2d updated ed. (New York: Wiley & Sons, 2000), ch. 15; George F. Ellis and William R. Stoeger, S.J., "Introduction to General Relativity and Cosmology," in *Quantum Cosmology and the Laws of Nature: Scientific Perspectives on Divine Action*, eds. Robert J. Russell, Nancey C. Murphy, and Chris J. Isham, Scientific Perspectives on Divine Action Series (Vatican City: Vatican Observatory Publications; Berkeley: Center for Theology and the Natural Sciences, 1993), 33–48. For a technical introduction, see Charles W. Misner, Kip S. Thorne, and John Archibald Wheeler, *Gravitation* (San Francisco: W. H. Freeman, 1973), part VI.

6.   The reference to space-time is actually based on Minkowski's geometrical inter-
        pretation of Einstein's SR. Although it is quite routine in scientific circles, it does
        not go entirely undisputed; cf. for example William Lane Craig, *Time and the
        Metaphysics of Relativity* (Dordrecht: Kluwer Academic Publishers, 2001).

7.   Thorne, Misner, and Wheeler, *Gravitation*.

8.   For a nontechnical introduction see Donald Goldsmith, *Einstein's Greatest Blun-
        der? The Cosmological Constant and Other Fudge Factors in the Physics of the Uni-
        verse* (Cambridge, MA: Harvard University Press, 1995), chs 10ff. For a more
        technical introduction see Chris J. Isham, "Creation of the Universe as a Quantum
        Process," in *Physics, Philosophy, and Theology: A Common Quest for Understand-
        ing*, eds. Robert J. Russell, William R. Stoeger, S.J., and George V. Coyne, S.J.
        (Vatican City State: Vatican Observatory Publications, 1988), 375–408; Edward
        W. and Michael S. Turner Kolb, *The Early Universe* (Reading: Addison-Wesley,
        1994).

9.   John D. Barrow and Frank J. Tipler, *The Anthropic Cosmological Principle* (Oxford:
        Clarendon, 1986), ch. 10; see also William R. Stoeger, S.J., "Scientific Accounts
        of Ultimate Catastrophes in Our Life-Bearing Universe," in *The End of the World
        and the Ends of God: Science and Theology on Eschatology*, eds. John Polkinghorne
        and Michael Welker (Harrisburg: Trinity Press International, 2000).

10.   If the universe is closed, then in 10^{12} years, the universe will have reached its
        maximum size and recollapse back to a singularity like the original hot Big Bang.

11.   Barrow and Tipler, *Cosmological Principle*, 648.

12.   Scholars who support an objective interpretation include Karl Barth, Raymond
        Brown, Gerald O'Collins, William Lane Craig, Stephen Davis, Wolfhart Pan-
        nenberg, Phem Perkins, Ted Peters, Janet Martin Soskice, Sandra Schneiders,
        and Richard Swinburne.

13.   Raymond E. Brown, *The Virginal Conception and Bodily Resurrection of Jesus*
        (New York: Paulist Press, 1973), 72.

14.   Scholars who support a subjective interpretation include Rudolf Bultmann, John
        Dominic Crossan, John Hick, Gordon Kaufman, Hans Küng, Sallie McFague,
        Willi Marxsen, Rosemary Radford Ruether, and Norman Perrin.

15.   Willi Marxsen, *The Resurrection of Jesus of Nazareth*, trans. Margaret Kohl (Phila-
        delphia: Fortress Press, 1970), 77, 156.

16.   Some scholars who refer to the resurrection as "objective" leave the question of
        physical and material continuity open-ended. For them, the objective interpreta-
        tion of the resurrection of Jesus is compatible with the possibility that his body
        suffered the same processes of decay that ours will after death; indeed it may even
        be seen as necessary (e.g., for his death to be like ours, etc.).

17.   Introductions to theology and science from a primarily Christian perspective
        include Robert John and Kirk Wegter-McNelly Russell, "Science," in *The Black-
        well Companion to Modern Theology*, ed. Gareth Jones (Oxford: Blackwell, 2004),
        512–56, Ian G. Barbour, *When Science Meets Religion: Enemies, Strangers or Part-*

 ners? (San Francisco: HarperSanFrancisco, 2000); Christopher Southgate, Celia Deane-Drummond, et al., eds., *God, Humanity and the Cosmos: A Textbook in Science and Religion* (Harrisburg: Trinity Press International, 1999); Ted Peters, "Theology and the Natural Sciences," in *The Modern Theologians: An Introduction to Christian Theology in the Twentieth Century,* 2d ed., ed. David F. Ford (Cambridge, MA: Blackwell, 1997), 649–68.

18. For an overview, see Robert John Russell, "Does 'The God Who Acts' Really Act? New Approaches to Divine Action in the Light of Science," *Theology Today* 51 (1997). Much of the research stems from the CTNS/VO series: Russell, Murphy, and Isham, eds., *Quantum Cosmology*; Robert John Russell, Nancey C. Murphy and Arthur R. Peacocke, eds., *Chaos and Complexity: Scientific Perspectives on Divine Action*, Scientific Perspectives on Divine Action Series (Vatican City State: Vatican Observatory Publications; Berkeley: Center for Theology and the Natural Sciences, 1995); Robert John Russell, William R. Stoeger, S.J., and Francisco J. Ayala, eds., *Evolutionary and Molecular Biology: Scientific Perspectives on Divine Action* (Vatican City State: Vatican Observatory Publications; Berkeley: Center for Theology and the Natural Sciences, 1998); Robert John Russell, Nancey Murphy, et al., eds., *Neuroscience and the Person: Scientific Perspectives on Divine Action* (Vatican City State: Vatican Observatory Publications; Berkeley: Center for Theology and the Natural Sciences, 1999); Robert John Russell, Philip Clayton, et al., eds., *Quantum Mechanics: Scientific Perspectives on Divine Action* (Vatican City State: Vatican Observatory Publications; Berkeley: Center for Theology and the Natural Sciences, 2001). Summaries of these papers can be found at http://www.ctns.org/books.html.

19. Stoeger, "Scientific Accounts of Ultimate Catastrophes in Our Life-Bearing Universe," 19–20.

20. Critical realists also claim that language is intrinsically metaphorical, and they defend a referential theory of truth warranted in terms of correspondence, coherence, and utility.

21. Ian G. Barbour, *Religion in an Age of Science*, Gifford Lectures: 1989–1990 (San Francisco: Harper & Row, 1990).

22. Arthur Peacocke, *Theology for a Scientific Age: Being and Becoming—Natural, Divine and Human*, enlarged ed. (Minneapolis: Fortress Press, 1993). See particularly fig. 3, p. 217, and the accompanying text.

23. Nancey Murphy, *Theology in the Age of Scientific Reasoning* (Ithaca: Cornell University Press, 1990).

24. Philip Clayton, *Explanation from Physics to Theology: An Essay in Rationality and Religion* (New Haven: Yale University Press, 1989).

25. John C. Polkinghorne, *The Faith of a Physicist: Reflections of a Bottom-up Thinker* (Princeton: Princeton University Press, 1994).

26. There are, of course, important differences between the methods of theology and the natural sciences, as Barbour and others stress carefully.

27.   One could do a more complicated diagram with physics, biology, and theology, for example, and one would need to include the influences of physics on both biology and theology, as well as the influences of biology on theology, etc.!

28.   This is a slight modification of Ellis's actual suggestion (personal communication).

29.   See, for example, the contradictory arguments about eschatology in John F. Haught, *God After Darwin: A Theology of Evolution* (Boulder: Westview, 2000), or the avoidance of the challenge of cosmology in Marjorie Hewitt Suchocki, *God, Christ, Church: A Practical Guide to Process Theology* (New York: Crossroad, 1982); Barbour, *Religion in an Age of Science*.

30.   John Polkinghorne, *The Way the World Is* (Grand Rapids: Eerdmans, 1983), ch. 8; Polkinghorne, *The Faith of a Physicist*, chs. 6, 9, esp. pp. 163–70; Polkinghorne, *Serious Talk: Science and Religion in Dialogue* (Valley Forge, PA: Trinity Press International, 1995), ch. 7.

31.   Polkinghorne, *The Faith of a Physicist*, 167.

32.   John Polkinghorne, "Eschatology: Some Questions and Some Insights from Science," in *The End of the World and the Ends of God: Science and Theology on Eschatology*, eds. John Polkinghorne and Michael Welker (Harrisburg: Trinity Press International, 2000), 29–30.

33.   Dante Alighieri, *The Divine Comedy*, trans. John Ciardi (New York: W. W. Norton, 1970), *The Paradiso*, canto XX, vs. 88–90.

34.   To view nature as created *ex nihilo* implies that the universe is contingent and rational, and these are two of the fundamental philosophical assumptions on which modern science is based. See for example Michael Foster, "The Christian Doctrine of Creation and the Rise of Modern Science," in *Creation: The Impact of an Idea*, ed. Daniel O'Connor and Francis Oakley (New York: Scribner's Sons, 1969); David C. Lindberg and Ronald L. Numbers, eds., *God and Nature: Historical Essays on the Encounter Between Christianity and Science* (Berkeley: University of California Press, 1986); Gary B. Deason, "Protestant Theology and the Rise of Modern Science: Criticism and Review of the Strong Thesis," *CTNS Bulletin* 6 (1986): 1–8; Christopher B. Kaiser, *Creation and the History of Science*, The History of Christian Theology Series, no. 3 (Grand Rapids: Eerdmans, 1991).

35.   Helge Kragh, *Cosmology and Controversy: The Historical Development of Two Theories of the Universe* (Princeton: Princeton University Press, 1996).

36.   Stoeger, "Contemporary Physics and the Ontological Status of the Laws of Nature," 209–34.

37.   Another way of making this case is to recognize that all scientific laws carry a *ceteris paribus* clause; that is, their predictions hold "all else being equal." But if God's regular action accounts for what we describe through the laws of nature, and if God acts in radically new ways to transform then world, then of course all else is not equal. I am grateful to Nancey Murphy for stressing this point to me (personal communication).

38.   It is crucial to note that the commitment to methodological naturalism does not carry any ontological implications about the existence/nonexistence of God (i.e., it is not "inherently atheistic").

39.   Since eschatology starts with the presupposition of God, it rules out reductive materialism and metaphysical naturalism. By taking on board natural science, other metaphysical options become unlikely candidates, including Platonic or Cartesian ontological dualism.

40.   See in particular the articles by Noreen Herzfeld, Nancey Murphy, Ted Peters, Jeffrey Schloss, and Michael Welker in the collection, *Resurrection: Theological and Scientific Assessments*, eds. Ted Peters, Robert John Russell, and Michael Welker (Grand Rapids, Eerdmans, 2002).

41.   I am grateful to Kirk Wegter-McNelly for suggesting the term *transcendental* here (personal communications).

42.   I agree with O'Collins's criticism that previous work on noninterventionist divine action did not deal with the problem of the resurrection. See Gerald O'Collins, S.J., "The Resurrection: The State of the Questions," in *The Resurrection: An Inter-disciplinary Symposium on the Resurrection of Jesus* (Oxford: Oxford University Press, 1997), 21, n. 52.

43.   O'Collins vs. Davis on "graced seeing" is a crucial issue here.

44.   Michael Welker offers a very imaginative formulation of the relation between the reign of God as already present and as apocalyptic in terms *eschatological comple-mentarity*. See his paper in Peters, et al., *Resurrection*.

45.   For a very creative development of this theme, see the paper by Nancey Murphy in Peters, et al., *Resurrection*.

46.   Caroline Walker Bynum, *The Resurrection of the Body in Western Christianity, 200–1336* (New York: Columbia University Press, 1995); Sandra Schneiders, "The Resurrection of Jesus and Christian Spirituality," in *Christian Resources of Hope* (Dublin: Columba, 1995), 81–114. See the article by Brian Daley in Peters, et al., *Resurrection*.

47.   Polkinghorne, "Eschatology."

48.   Russell, "The God Who Infinitely Transcends Infinity" (Berkeley: University of California Press 1996).

49.   For a helpful overview of Trinitarian theologians on the problem of "time and eternity," see Peters, *God as Trinity* and Russell, "Time in Eternity," *Dialog* 39 (2000). There are, of course, fundamental problems in contemporary discussions of the doctrine of the Trinity. These include the source of the doctrine (is it ana-lytic or synthetic?), the meaning of the divine persons, their principle of unity, the relation of the economic and the immanent Trinity, and so on. Here I am not attempting to adjudicate between these issues. I am simply lifting up some themes which, arguably, most Trinitarian formulations have in common in order to suggest ways to start the conversation with science.

50.   In brief, Wolfhart Pannenberg claims that duration was part of the biblical and early Western understanding of time, that it was lost in Augustine's separation of subjective and objective time, and that modern physics inherited the latter, durationless view. He then points to recent philosophy where attempts have been made to view physical time in terms of duration, as in the writings of Bergson, Heidegger, and Whitehead, whose roots lie in part in Christian theology. See Wolfhart Pannenberg, "Theological Questions to Scientists," in *The Sciences and Theology in the Twentieth Century*, ed. A. R. Peacocke (Notre Dame: University of Notre Dame Press, 1981); idem, *Metaphysics and the Idea of God* (Grand Rapids: Eerdmans, 1990); idem, *Systematic Theology*, trans. G. W. Bromiley (Grand Rapids: Eerdmans, 1991), 1; idem, *Toward a Theology of Nature: Essays on Science and Faith*, ed. Ted Peters (Louisville: Westminster/John Knox Press, 1993).

51.   The focus on "time and eternity" is particularly appropriate for our task here since most contemporary theologians root their eschatology within the framework of *creatio ex nihilo*. Thus much of what they claim about time and eternity applies to the universe as the creation, and not only as it will be as new creation.

52.   Scholars such as William Lane Craig insist that we should return to Einstein's original classical framework of matter moving in three-dimensional Euclidean space with time as an independent parameter. See Craig, *Time and the Metaphysics of Relativity*.

53.   J. R. Lucas, *The Future: An Essay on God, Temporality, and Truth* (Oxford/New York: Blackwell, 1989).

54.   Ironically, its most crucial technical problem, "the measurement problem," requires the assumption of a universal present and a flowing time.

55.   A possible mathematical model for the concept of copresence is a non-Hausdorff manifold in which distinct events are not separable topologically.

56.   J. A. Wheeler and R. P. Feynman, "Interaction with the Absorber as the Mechanism of Radiation," *Rev. Mod. Phys.* 17 (1945): 156; Wheeler and Feynman, "Classical Electrodynamics in Terms of Direct Interparticle Action," *Rev. Mod. Phys.* 21 (1949): 424.

57.   Fred Hoyle and Jayant V. Narlikar, *Action at a Distance in Physics and Cosmology* (San Francisco: W. H. Freeman, 1974).

58.   Fred Hoyle, Geoffrey Burbidge, and Jayant V. Narlikar, *A Different Approach to Cosmology: From a Static Universe Through the Big Bang Towards Reality* (Cambridge: Cambridge University Press, 2000).

59.   For a readable account see Goldsmith, *Einstein's Greatest Blunder?* For technical arguments see Lawrence M. Krauss, "The End of the Age Problem, and the Case for a Cosmological Constant Revisited," *CWRU-P6-97/CERN-Th-97/122/Astro-Ph/9706227 Preprint* (1997).

60.   George L. Murphy, "Hints from Science for Eschatology—and Vice Versa," in *The Last Things: Biblical & Theological Perspectives on Eschatology*, ed. Carl E. Braaten and Robert W. Jenson (Grand Rapids: Eerdmans, 2002), 157–60.

61.   For a technical discussion see George F. R Ellis, J. Hwang, and M. Bruni, *Phys. Rev. D 40* (1989): 1819–26; Stephen Hawking and George F. R. Ellis, *The Large Scale Structure of Space-Time*, Cambridge Monographs on Mathematical Physics Series (Cambridge: Cambridge University Press, 1973), 88–96.

62.   K. Gödel, "An Example of a New Type of Cosmological Solution of Einstein's Field Equations of Gravitation," *Rev. Mod. Phys.* 21 (1949): 447–50.

63.   For somewhat accessible background material see Misner and Wheeler, *Gravitation*, part VII.

64.   Misner and Wheeler, *Gravitation*, ch. 34. For technical material, see Hawking and Ellis, *Space-Time*. The discussion of these topics in the context of quantum cosmology is much more complex. See N. D. Birrell and P. C. W. Davies, *Quantum Fields in Curved Space* (Cambridge: Cambridge University Press, 1982); Isham, "Creation of the Universe"; Isham, *Lectures on Quantum Theory: Mathematical and Structural Foundations* (London: Imperial College Press, 1995).

Chapter Seven

1.   For an extensive discussion, see Hugh Ross, "Extra-Dimensionality and the New Creation," *Beyond the Cosmos: What Recent Discoveries in Astrophysics Reveal about the Glory and Love of God*, rev. ed. (Colorado Springs: NavPress, 1999), 217–28.

2.   Hugh Ross, *Beyond the Cosmos*, 73–79.

3.   Genesis 1:1, Psalm 148:5, Isaiah 45:18, Hebrews 11:3, Genesis 2:3, Isaiah 40:26, John 1:3, Genesis 2:4, Isaiah 42:5, Colossians 1:15-17.

4.   Hugh Ross and John Rea, "Big Bang—The Bible Taught It First!" *Facts for Faith* 3 (2000): 28–29.

5.   As Bruce Waltke explained in his Kenneth S. Kantzer Lectures in Systematic Theology given January 8–10, 1991, at Trinity Evangelical Divinity School, Deerfield, IL, the Hebrew words *shāmayim* and *'eres*, when placed together, form a compound word that, like the English compound word butterfly, takes on a meaning of its own.

6.   Hugh Ross, *The Creator and the Cosmos: How the Greatest Scientific Discoveries of the Century Reveal God*, 3rd ed. (Colorado Springs: NavPress, 2001), 102.

7.   Roger Penrose, *Shadows of the Mind: A Search for the Missing Science of Consciousness* (New York: Oxford University Press, 1994), 230: "This makes Einstein's general relativity, in this particular sense, the most accurately tested theory known to science!"

8.   Stephen W. Hawking and Roger Penrose, "The Singularities of Gravitational Collapse and Cosmology," *Proceedings of the Royal Society of London* A 314, no. 1519 (1970): 529–48, abstract available from http://adsabs.harvard.edu/cgi-bin/nph-bib_query?bibcode=1970RSPSA.314..529H&db_key=AST&high=3f1c1fc06718016 (accessed January 23, 2004).

9.   Job 9:8, Isaiah 42:5, Isaiah 48:13, Jeremiah 51:15, Psalm 104:2, Isaiah 44:24, Isaiah 51:13, Zechariah 12:1, Isaiah 40:22, Isaiah 45:12, Jeremiah 10:12.

10.   The four expanding dimensions are length, width, height, and time.

11.   Lawrence M. Krauss, "The End of the Age Problem, and the Case for a Cosmological Constant Revisited," *Astrophysical Journal* 501 (1998): 461.

12.   Krauss, "End of the Age Problem," 465.

13.   Hugh Ross, "Fine-Tuning for Life in the Universe," *Reasons To Believe*, available from http://www.reasons.org/resources/apologetics/design_evidences/200205 02_universe_design.shtml?main (accessed February 26, 2004).

14.   Genesis 1:1, 2; Ecclesiastes 1:3-15; Romans 8:18-23; Genesis 2:5-6; Jeremiah 33:25; Revelation 21:1–22:5.

15.   Lawrence M. Krauss and Glenn D. Starkman, "Life, the Universe, and Nothing: Life and Death in an Ever-Expanding Universe," *Astrophysical Journal* 531 (2000): 22–30.

16.   Ruth A. Daly and S. G. Djorgovski, "A Model-Independent Determination of the Expansion and Acceleration Rates of the Universe as a Function of Redshift and Constraints on Dark Energy," *Astrophysical Journal* 597 (2003): 9–20.

17.   Einstein's theory of special relativity states that nothing can be accelerated past the velocity of light in the absence of the space energy density factor. There is no limit to how rapidly the space energy density (also called dark energy) factor can accelerate the growth of the surface of the universe.

18.   Krauss and Starkman "Life in the Universe," 29.

19.   Matthew 24:35; Mark 13:31; Luke 21:33.

20.   Yu N. Mushurov and I. A. Zenina, "Yes, the Sun is Located Near the Corotation Circle," *Astronomy & Astrophysics* 341 (1999): 81–85; J. R. D. Lépine, I. A. Acharova, and Yu. N. Mishurov, "Corotation, Stellar Wandering, and Fine Structure of the Galactic Abundance Pattern," *Astrophysical Journal* 589 (2003): 210–16; Peter Hoppe et al., "Type II Supernova Matter in a Silicon Carbide Grain from the Murchison Meteorite," *Science* 272 (1996): 1314–16. Yu N. Mishurov, J. R. D. Lepine, and I. A. Acharova, "Corotation: Its Influence on the Chemical Abundance Pattern of the Galaxy," *Astrophysical Journal Letters* 571 (2003): 113–15.

21.   Psalm 19:1-4; Psalm 50:6; Psalm 97:6.

22.   Robert H. Dicke, "Dirac's Cosmology and Mach's Principle," *Nature* 192 (1961): 440.

23.   Hugh Ross, *Creator and the Cosmos*, 96.

24.   Ibid., 157–60.

25.   Freeman Dyson, "Energy in the Universe," *Scientific American* 224 (1971): 59.

26.   Paul Davies, *The Cosmic Blueprint: New Discoveries in Nature's Creative Ability to Order the Universe* (New York: Simon & Schuster/Touchstone, 1988), 203.

27.   Brandon Carter, "The Anthropic Principle and Its Implications for Biological Evolution," *Philosophical Transactions of the Royal Society* A 370 (1983): 347–60.

28. S. Sahijpal et al., "A Stellar Origin for the Short-Lived Nuclides in the Early Solar System," *Nature* 391 (1998): 559–61; G. J. Wasserburg, R. Gallino, and M. Busso, "A Test of the Supernova Trigger Hypothesis with ^{60}Fe and ^{26}Al," *Astrophysical Journal Letters* 500 (1998): L189–93; Peter Hoppe et al., "Type II Supernova Matter in a Silicon Carbide Grain from the Murchison Meteorite," *Science* 272 (1996): 1314–16.

29. Neil F. Comins, *What if the Moon Didn't Exist? Voyages to Earths That Might Have Been* (New York: HarperCollins: 1993), 2–8; H. E. Newsom and S. R. Taylor, "Geochemical Implications of the Formation of the Moon by a Single Giant Impact," *Nature* 338 (1989): 29–34; Jack J. Lissauer, "It's Not Easy to Make the Moon," *Nature* 389 (1997): 327–28; Sigeru Ida, Robin M. Canup, and Glen R. Stewart, "Lunar Accretion from an Impact-Generated Disk," *Nature* 389 (1997): 353–57.

30. Brandon Carter, "*Anthropic Principle*," 347–60; also see John D. Barrow and Frank J. Tipler, *The Anthropic Cosmological Principle* (New York: Oxford University Press, 1986), 566.

31. Barrow and Tipler, *Anthropic Cosmological Principle*, 556–70.

32. Juliana Sackmann and Arnold I. Boothroyd, "Our Sun v. A Bright Young Sun Consistent with Helioseismology and Warm Temperatures on Ancient Earth and Mars," *Astrophysical Journal* 583 (2003): 1024–39.

33. For a discussion of this process, see Hugh Ross, "The Faint Sun Paradox," *Facts for Faith* 3 (2002): 28–33; also see "The Faint Sun Paradox," Reasons To Believe, available from http://www.reasons.org/resources/fff/2002issue10/index.shtml?main#faint_sun (accessed February 27, 2004).

34. Robert G. Brakenridge, "Terrestrial Paleoenvironmental Effects of a Late Quaternary-Age Supernova," *Icarus* 46 (1981): 81–93.

35. James F. Crow, "The Odds of Losing at Genetic Roulette," *Nature* 397 (1999): 293.

36. Hugh Ross, "Anthropic Principle: A Precise Plan for Humanity," *Facts for Faith*, quarter 1 (2002): 24–30; also see "A Precise Plan for Humanity," Reasons To Believe, available from http://www.reasons.org/resources/fff/2002issue08/index.shtml?main#a_precise_plan (accessed February 27, 2004).

37. The Genesis text in which God describes the greatly increased pain of childbearing may refer not just to physical pain but also to a pain beyond the physical realm—a pain that impacts mothers and fathers. That pain comes from knowing in advance that one's precious children have free will and a sin nature that will surely be expressed and that one's children will unavoidably be impacted by others' sin. This is a dreadful realization, a pang more severe, according to my wife, than any bodily birth pang.

38. C. S. Lewis, *The Last Battle* (New York: Macmillan, 1956), 173–74.

Chapter Eight

1.   For more information on the current activity surrounding artificial intelligence, see S. Jennifer Leat, "Artificial Intelligence Researcher Seeks Silicon Soul," *Research News & Opportunities in Science and Theology*, 3 (2002): 7, 26; Paul Mullin, "Can the Image of God Ever Be Artificial?" *Research News & Opportunities in Science and Theology* 3 (2003): 7, 27; Anne Foerst, "Artificial Intelligence Gives New Meaning to Personhood," *Research News & Opportunities in Science and Religion* 2 (2002): 22, 32; Heather Wax, "Humble Checkers Program Raises AI Issues for Decades," *Research News & Opportunities in Science and Theology* 4 (2003): 35; Paul Davies, "Quantum Computing: A Key to Unlocking Reality?" *Science & Spirit* 11 (2000): 22–23, 41, 50.

2.   Arthur Eddington, *The Expanding Universe* (New York: Cambridge University Press, 1987).

3.   Samuel Alexander, *Space, Time and Deity* (New York: Dover, 1966). First published in 1920, this book represents the body of Alexander's Gifford Lectures of 1916–1918.

4.   Rudolph Bultmann, *Theology of the New Testament*, 2 vols., trans. Kendrick Grobel (New York: Scribner's Sons, 1951–1955).

5.   Eddington, *Expanding Universe*, 13.

6.   C. S. Lewis, *Miracles* (London: Geoffrey Bles, 1947).

7.   Edwin A. Abbot, *Flatland*. All of Einstein's students were required to read *Flatland* because of its creative picture of modern cosmology.

8.   Ben Patterson served as chaplain for the 2002 C. S. Lewis Summer Institute.

9.   Lewis discusses time in his contemplation of prayer found in *Mere Christianity*, (New York: Macmillan, 1952), 130–33.

10.   The theme verse for the 2002 C. S. Lewis Summer Institute was Ecclesiasties 3:1-11.

11.   Nietzsche explores his concept of eternal recurrence throughout *The Will to Power* (New York: Random, 1967).

12.   Hawking presents his alternative to the Big Bang in *A Brief History of Time* (New York: Bantam Books, 1988).

13.   See Teilhard de Chardin, *The Phenomenon of Man* (New York: Harper & Row, 1965).

14.   T. S. Eliot, "The Hollow Men," in *Collected Poems, 1909–1935* (London: Faber & Faber, 1936), 90.

15.   Nigel Goodwin, founder of Genesis Arts Trust of London, serves on the board of the C. S. Lewis Foundation.

16.   Oscar Cullmann, *Christ and Time*, trans. Floyd V. Filson (London: SCM Press, 1951), 141.

17.   Julia Ward Howe, "The Battle Hymn of the Republic," public domain.

Conclusion

1.   Greg Boyd makes this point in *Is God to Blame?: Beyond Pat Answers to the Problem of Pain* (Downers Grove: InterVaristy Press, 2003).

2.   This issue formed the basis for the plenary sessions of the 2001 annual meeting of the Evangelical Theological Society. The plenary addresses were published by the society's journal, *Journal of the Evangelical Theological Society* 45 (2002).

3.   Julian Barbour, *The End of Time* (New York: Oxford University Press, 1999).

List of Contributors

TONY CAMPOLO is professor emeritus of sociology at Eastern University in St. David's, Pennsylvania. For ten years he was on the faculty at the University of Pennsylvania. He did his undergraduate work at Eastern College and earned his Ph.D. from Temple University. Campolo also serves as an associate pastor of the Mount Carmel Baptist Church in West Philadelphia, and serves as an associate for international ministries of American Baptist Churches. He is the founder of the Evangelical Association for the Promotion of Education (EAPE) and the author of many popular books, including *A Reasonable Faith, It's Friday but Sunday's Coming, Let Me Tell You a Story, Carpe Diem, Which Jesus?,* and *Following Jesus Without Embarrassing God.*

WILLIAM LANE CRAIG is research professor of philosophy at Talbot School of Theology in La Mirada, California. Dr. Craig pursued his undergraduate studies at Wheaton College (B.A. 1971) and graduate studies at Trinity

Evangelical Divinity School (M.A. 1974; M.A. 1975), the University of Birmingham (England) (Ph.D. 1977), and the University of Munich (Germany) (D. Theol. 1984). From 1980–1986 he taught Philosophy of Religion at Trinity. In 1987 Craig moved to Brussels, Belgium, where he pursued research at the University of Louvain until 1994. He is the author of *God and Time, Time and Eternity, God, Time and Eternity,* and *Time and the Metaphysics of Relativity.*

TIMOTHY GEORGE serves as professor of church history and dean of the Beeson Divinity School of Samford University in Birmingham, Alabama. George holds the M.Div. from Harvard Divinity School and the Ph.D. from Harvard University. He has served as pastor or interim pastor of ten Baptist churches. He also serves as executive editor of *Christianity Today* and on the editorial board of *Books & Culture.* A popular lecturer, George is a frequent speaker at universities and seminaries, conferences, and churches. He serves on many boards, including the Center for Catholic and Evangelical Theology, Prison Fellowship Ministries, and Wheaton College. He is the author of many works on church history, theology, and biblical studies, including *John Robinson and the English Separatist Tradition, The Theology of the Reformers, Galatians* (New American Commentary), *Is the Father of Jesus the God of Muhammad?,* and *A Mighty Long Journey: Reflections of Racial Reconciliation,* among many others.

J. STANLEY MATTSON holds the M.A. from the University of Wisconsin and the Ph.D. in American social and cultural history from the University of North Carolina at Chapel Hill. He taught history at North Carolina State University and Gordon College before accepting leadership posts as headmaster of The Master's School in West Simsbury, Connecticut and director of corporate and foundation relations at the University of Redlands in Redlands, California. He resigned that position in 1986 to found the C. S. Lewis Foundation which he has led ever since.

HARRY LEE POE serves as Charles Colson Professor of Faith and Culture at Union University in Jackson, Tennessee. He has written several books and numerous articles on how the gospel intersects with culture; including *Christianity in the Academy, The Gospel and Its Meaning, Christian Witness in a Postmodern World,* and *See No Evil: The Existence of Sin in an Age of Relativism.* Poe wrote *The Designer Universe* and *Science and Faith: An Evangelical Dia-*

logue with Jimmy H. Davis with whom he collaborates on science and religion issues. Poe also serves as program director for the triennial C. S. Lewis Summer Institute in Oxford and Cambridge and as president of the Edgar Allan Poe Museum of Richmond, Virginia.

SIR JOHN POLKINGHORNE is an Anglican priest, president emeritus of Queens' College, Cambridge University, a member of the Royal Society, and former professor of mathematical physics at Cambridge. Polkinghorne resigned his chair in physics to study for the Anglican priesthood. After completing his theological studies and serving as a parish priest, he returned to Cambridge. During the same time period, he wrote a series of books on the compatibility of religion and science. These include *Science and Creation*, and most recently, *Science and Providence*, and his Gifford Lectures, "The Faith of a Physicist." Sir John was the recipient of the 2002 Templeton Prize.

HUGH ROSS holds an undergraduate degree in physics from the University of British Columbia and the M.S. and Ph.D. in astronomy from the University of Toronto. He served as research fellow in radio astronomy at the California Institute of Technology, 1973–1978. Because of the role that science played in his conversion to faith in the God of the Bible, Ross has undertaken an apologetics ministry to help people understand the relationship between modern science and biblical faith. He is the founder and director of Reasons to Believe and hosts a television program by the same name. His Web site may be consulted at *www.reasons.org*. Ross also serves as associate minister, Sierra Madre Congregational Church. He is the author of many technical articles as well as popular books, including *Creation and Time*, *The Genesis Question*, *The Creator and the Cosmos*, *Beyond the Cosmos*, and *A Matter of Days*.

ROBERT JOHN RUSSELL serves as professor of theology and science in residence at the Graduate Theological Union in Berkeley, California. Russell was educated in both physics and theology. After graduating in physics from Stanford (1968), Russell earned the M.S. in physics at the University of California, Los Angeles, in 1970. He went on to earn the B.D. and M.A. in theology at Pacific School of Religion in 1972. He was awarded the Ph.D. in physics by the University of California, Santa Cruz in 1978. He was ordained to the ministry of higher education by the United Church of Christ (Congregational) in 1972. He was professor of physics at Carleton College (MN) for several years before moving to the Graduate Theological Union in California where he serves as professor of theology and science in residence. He is the founder and director of the Center for Theology and Natural Sciences in Berkeley. He has also served as general editor of a series of research publications on scientific perspectives on divine action. He is the coeditor of a number of books on science and religion, including *Physics, Philosophy and Theology: A Common Quest for Knowledge, John Paul II on Religion and Science: The New View from Rome, Quantum Cosmology and thee Laws of Nature: Scientific Perspectives on Divine Action,* and *Resurrection: Theological and Scientific Perspectives.*

Dr. Russell Stannard, OBE, is emeritus professor of physics of the Open University in the United Kingdom where he served as head of the physics department for many years. He also has been named a fellow of University College London. He is a recipient of the Bragg Medal and Prize of the Institute of Physics and gained international recognition as a regular broadcaster on BBC radio and television in matters relating to science. Stannard is a lay reader in the Church of England and serves as a trustee of the John Templeton Foundation. He is the author of many books including *Science and the Renewal of Belief, Grounds for Reasonable Belief, Doing Away With God?,* and *Science and Wonders.* His Gifford Lectures have been published as *The God Experiment.* In addition to his scholarly work, he has written eleven books for children on issues related to science.

Index